Ann Lingard spent her childhood in C
ing in various places including Cambri
moved to a Cumbrian small-holding
Having left academia and scientific research to write and broadcast, she
has published six novels – including *Seaside Pleasures* and *The Embalmer's
Book of Recipes* – and several short stories (see www.eliotandentropy.
wordpress.co.uk), and many articles about the countryside and shore.
She blogs about the Solway at www.solwayshorewalker.co.uk.

Praise for *The Fresh and the Salt*

'Like a hungry gull, Ann Lingard explores her beloved Solway shore-
line for every living detail that catches her eye . . . She lets no detail
escape her notice and in so doing has created a portrait of this
nation-cleaving water that is as broad and deep as the estuary itself. A
wonderful addition to the literature of place.'

Mark Cocker, author and naturalist

'Beautiful, intensely visual prose, born from deep intimacy with subtle
borderlands: land and sea, England and Scotland, people and environments. Lingard expertly probes the margins for their hidden riches.'

David Gange, author of *The Frayed Atlantic Edge*

'This is deep and beautiful natural history writing rather than nature
writing.'

Fergus Collins, *Countryfile* magazine

'Catching the poetic in the scientific, and steeped in environmental
histories of the area, Ann addresses saltmarsh and mudflat, song and
painting, mudshrimp and stonemason with the same curiosity. From
the Newton Arlosh saltmarshes via saltworks and wildfowling, to the

wetlands of Caerlaverock, this is a kaleidoscopic portrait of the borders of the land.'

Will Smith, *Cumbria Life* magazine

'A natural history in the richest sense of the term . . . I admire the clarity that Lingard brings to the relationships between creatures and places, people and places, and indeed to the relationship between one place and another. It takes a certain delicacy of language to tease this out.'

Professor Isaac Land, www.porttowns.port.ac.uk

'A cabinet of curiosities, an engaging portrait not only of place but of a particular way of seeing; one that sets out to investigate and celebrate much more than that which lies merely upon the surface.'

Karen Lloyd, www.caughtbytheriver.net

'Its comprehensive quality and informed view mean that the book will serve, not just as a book about Solway, but also as an exemplar for readers' experience of their own estuary ecosystem, serving to direct attention to diverse aspects of ecological relations of the whole.'

Peter Reason, www.shinynewbooks.co.uk

The Fresh and the Salt

THE STORY OF THE SOLWAY

ANN LINGARD

BIRLINN

This edition first published in 2024 by
Birlinn Limited
West Newington House
10 Newington Road
Edinburgh
EH9 1QS

Copyright © Ann Lingard 2020

First published in hardback in 2020

ISBN 978 1 78027 849 0

British Library Cataloguing in Publication Data
A catalogue record for this book is available from the British Library.
Designed and typeset by Biblichor, Edinburgh

Typeset by Biblichor Ltd, Edinburgh
Printed and bound by Clays Ltd, Elcograf, S.p.A

Brown clouds are blown against the bright fells
Like celtic psalms from drowned western isles.
The slow rain falls like memory
And floods the becks and flows to the sea,
And there on the coast of Cumberland mingle
The fresh and the salt, the cinders and the shingle.

From Norman Nicholson's 'Five Rivers' (1944)[1]

Contents

List of Plates

List of Plates

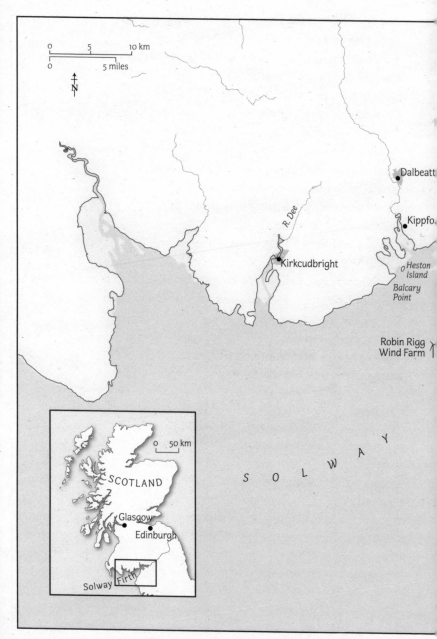

The Solway Firth

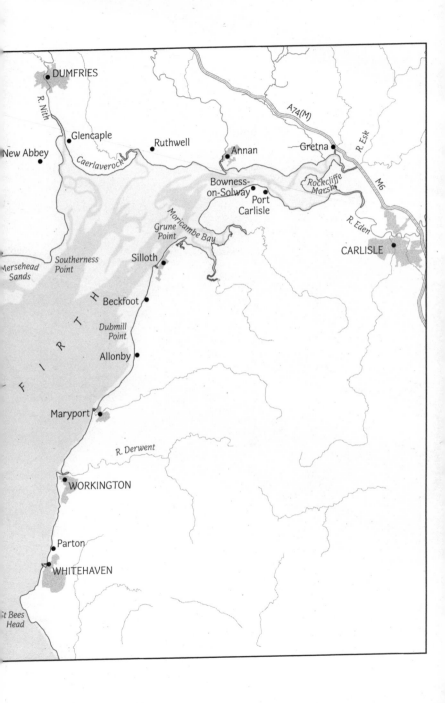

Introduction

We are flying slowly, low over Rockcliffe Marsh at the head of the Solway Firth. On each side of the vast saltmarsh the beds of the rivers Eden and Esk are knitted with patterns of light and dark that record their turbulent flow. The sunlight slanting through the clouds shifts shapes and colours across the land and water – dark greens, silver-satin and bright blue. Mudflats are patches of ochre and burnt sienna, sandbanks are grey and gold, and water glimmers and flashes from creeks and pools, changing with the angle of our flight.

I can enjoy it now. I can even lean out over the side confident that my seatbelt will hold me and that I won't fall, tumbling and spinning, to disappear like a tossed pebble in the sticky mud that lies below. I can even speak again without my tongue clacking dryly against my palate. Earlier, at Carlisle Airport, I had been fitted with flying suit, scarf and helmet; hung my camera's lanyard round my neck; found my gloves (it would be cold); discarded my notebook (it would be windy). Andrew Lysser[1] had explained some of the mechanics of the gyroplane to me and had given me a few important tips about communication and emergency procedures (this wasn't the standard EasyJet briefing, so I *listened*); and then we had walked outside to look at the gyroplane. The open-sided flying-machine, with rotors above and behind that looked ridiculously flimsy and a motor that sounded like a hornet, was rather shocking. Getting in required some flexibility. The control tower gave Andrew clearance and we motored across to the runway, then accelerated. It seemed impossible that we could gain enough lift to leave the ground, but Andrew's voice through my earphones told me, 'We'll take off when the top rotor reaches two hundred revs per minute.' As I looked out to my right, I saw a gap open between the shadow of the wheels and the runway and I realised we had become airborne, smoothly. I felt very exposed; there was nothing between me and 'outside'. Andrew talked quietly to me over the radio – his calm voice could persuade a potential suicide to step back from the brink – and as he pointed out the recent alterations to the airport perimeter, a mansion belonging to someone well-known and, soon, the huge expanse of Carlisle's railway marshalling yards, it became easier to enjoy the experience. We were seeing places which, screened at ground level, I hadn't even known existed. 'You see those buildings down there, with the trees all around? That's a naturist colony!' In chilly north-west Cumbria! By then I was nearly relaxed enough to laugh.

I had wanted this different perspective on the Solway because it's not easy from shore level to appreciate the relationships between the bays and ports and rivers of the Firth's two coasts. The Solway Firth is a crooked finger of water reaching far inland from the fist of the Irish Sea, apparently prising Scotland away from England: there is an invisible border down the midline of the Firth, yet the sea also unites the two countries as a means of passage. The word 'firth' – or more commonly in Victorian times, 'frith' – originates from a Norse word which also gave rise to 'fjord', and one definition is adamant that this differs from 'estuary'. But at least twelve rivers, becks and burns flow into the Solway, and there is no question that the Upper Solway is a major estuary, into which open many smaller estuarine mouths. Even at its innermost tip, where it touches the Border at Gretna, the Firth experiences a tidal flow. Indeed, the Solway is famous for the range and speed of its tides, which during the biggest spring tides may have a range of nine to ten metres.[2] In its entirety the Firth stretches from the Mull of Galloway in Scotland across to St Bees Head in Cumbria, but for many people, including me, the 'Firth' means the Upper Solway, the land and sea to the east of a line between St Bees and the Mull of Ross west of Kirkcudbright (pronounced Kerr-*coo*-bree).

Zoom out to a much higher view than from a gyroplane, to the stitched-together satellite images on Google.[3] There is Ireland's east coast and the Irish Sea, the Isle of Man, and the whole of the Solway Firth stretching inland to the porous border between the nations of Scotland and England. Increase the magnification and slide the view to see the Solway in greater detail. There are the river channels, their beds braided by variations in their flow. The graded colours of sand and mud and marsh and the khaki channels of the rivers have been revealed by the ebbing tide – the amalgamated satellite images must all have been grabbed at a similar state of time and tide. From both Scotland and England the branches of wriggling creeks anastomose through saltmarshes like the ducts in a kidney, and estuarine mudflats bleed water into their rivers. But these satellite views also show the glacial origins of the landscape – the brownish-red patches of the lowland raised bogs that grow upwards in wet glaciated hollows; the farmsteads built on drumlins above the flood-prone land; the relict glacial dumps of rocks, dark freckled patches along the shore. It is only 10,000 years – mere minutes on the clock-face of geological time – since the glaciers melted and freed the land and sea and rivers to begin their dance. Around the malleable edges of the Upper Solway we are only scraping the surface of the 'deep time' that lies below its seabed.

I could stare at the satellite view for hours, dragging the image from west to east, from north to south, to find where I had walked, squelched, waded, slipped and scrambled along the shores on both sides of the Firth: foot-fall words, for exploring on the ground.

Drop back down to gyroplane height, to fly south-west over Burgh (pronounced Bruff) Marsh, where startled cattle flee, rocking, their tails erect. Their well-trodden paths meander across the saltmarsh, muddying the ground around the uninspiring monument to King Edward I, the 'Hammer of the Scots', who died here of dysentery while trying to cross to Scotland in 1307. Nearby is Port Carlisle, its former industrial splendour now decayed, its derelict coaling wharf a tumble of red sandstone. A line of wooden stumps is all that remains of a former steamer pier where emigrants, hopeful, sad or anxious, embarked for

Liverpool and onwards to America, and the line of terraced houses that was built to service the port has an elegant simplicity. At Bowness-on-Solway houses fringe the road that curves up and away from the mudflats and onto the rise at the end of Hadrian's Wall. The mud has been etched by the currents that sweep round on the English side, and the estuary here is shallow – a wath or crossing-point for humans and their cattle. Sandy shoals cast off the ebbing tide. We can see gulls resting on the water's surface, and Andrew tells me that he has looked down on swimming salmon. Here, between Bowness and Annan, the two countries are very close, and at low tide fishermen haaf-netting for salmon and sea trout wade between the shores. In 1934 the only physical link across the Firth, the railway viaduct, was demolished.[4] Andrew pilots the gyroplane lower so that we can look at the red sandstone stubs of the viaduct's embankments jutting out from either side of the Firth amongst the saltmarshes and exposed mudflats. Other firths and major estuaries are bridged – would the relationship between Cumbria and Dumfries & Galloway have developed differently if some kind of bridge still existed?

We gain some height again, and now the view as we continue to head west along the length and width of the Solway is breathtakingly beautiful. Water glimmers in all directions in wide sweeps and layers of colour; curving green channels; the alternating dark and light bands of underwater sand waves; golden-brown banks of sand shouldering above the glitter. Sheets of light are stretched over the water and billow over mudflats. The estuarine mouths of the Nith and Lochar Water to the north are wide. When the tide returns it will be no great distance to voyage from port to port or to the Irish Sea. We examine the edges further west along the English shore: the difficult entry to Silloth's Port; the rocky scaurs and named boulders of the Allonby shore; the overgrown saltpans and foundations of the Roman mile-fortlet at Crosscanonby. But a grey curtain is being drawn across the south, obliterating the Lake District Fells, the fields and woods, the outer Firth, the Scottish hills. The rain is driving north and east towards us. If we can't find a 'hole' through it or fly around it we will have to turn round for, alarmingly, it seems that too much rain damages the rotor tips, and without a lid over us we would get very wet. The distant wall of rain looks solid and wide and low.

As we turn to the north-east the great disc of mudflats and sand in Moricambe Bay is exposed and pale beneath us, and scribbled with shining water.[5] Andrew drops the gyroplane in a loop to show me the lumpy metal engine of a Lockheed Hudson bomber which protrudes sufficiently from its grave to provide a perch for a single gull.[6] Cormorants sitting motionless with outstretched wings on wooden posts are flurried into the air by our noisy clatter overhead. The wide saltmarsh that fringes the bay is dotted with grazing cattle, and two little egrets flap slowly along the margin, improbably white against the already darkening green. We circle, hoping to find the edge of the cloud, but rain is spattering against the windshields. We retreat, heading back above the richly coloured peatbogs of the Mosses towards Carlisle: re-entering bright light, warmth and colour.

*

Those wooden posts out in the mud of Moricambe Bay were the remnants of a World War II practice target. The targets for bombing or firing practice needed to be sited well away from populated areas, so constructing them out in the Firth itself was an obvious choice. Pilots and navigators were pointed in the right direction, literally, by large concrete arrows that were embedded in the dunes and cliffs. There are arrows on the Scottish coast, the concreted area at Carsethorn now enterprisingly used as hard-standing for a clamp for storing silage. The posts of the target out on the Mersehead sands are enticingly visible from the sweeping sands of Sandyhills Bay, but the distance is deceptive. During the war, RAF Dumfries was a large and busy airport, with maintenance units and a training school for bombers and gunners; aircraft would have been arriving for repair or modification every day, and others would have been heading out over the Firth to practise firing and bombing.[7] The air-crews might have glanced down to the beaches and glittering water beneath them and seen too the sandbanks of Robin Rigg and Barnhourie revealed by the falling tide. Perhaps in the summers they would even have picked out a basking shark, a large, slow-moving shape so clearly described by Kathleen Jamie:

> . . . its ore-
> heavy body and head –
> the tail fin measuring back,
> forth, like a haunted door – [8]

Back at ground level, I revisit the two arrows that remain near Mawbray on the English side, their edges still sharp and their surfaces scattered with rabbit droppings. It is late July and the sandy vegetation around the nearby shallow pond is heaving with natterjack toadlets, crawling and pushing their way through the stalks, perilously oblivious to the paws of dogs and the large feet of humans. I kneel to watch these tiny fingernail-sized creatures, their backs brownish-black and warty, the characteristic yellow stripe already developing, as they struggle through the forest of vegetation. After feeding and growing throughout the summer and autumn, they will burrow into the sand to over-winter. The spring mating chorus of the adult males is notoriously loud, echoing around the dunes and saltmarshes.[9] That they prefer to breed in shallow, ephemeral ponds – where the water is warmer and few predators, like the fiercely carnivorous dragonfly larvae, can live – does not help their survival, but, with practical help from various conservation organisations, they are hanging on, and the Scottish saltmarshes of Mersehead and Caerlaverock, and the Cumbrian dunes are home to the majority of the UK's natterjack population. Other populations are found on the Merseyside coast amongst the Sefton sand dunes, and also on the south coast of Ireland. According to legend, the Irish natterjacks arrived there as shipwreck survivors from an eighteenth-century sailing ship from Liverpool that had taken on Sefton sand as ballast.

Another sign that the season is changing is the arrival of the geese on the Solway: pinkfooted geese from Greenland and Iceland, *wink-wink-wink*-ing overhead, and the delicate black-and-white barnacle geese from Svalbard. The talkative, wavering Vs of birds reach us in late autumn and settle in undulating sheets on the fields and marshes. People look forward to their return and comment too on the flocks of oystercatchers gathering at the sea's edge, the binary dark-or-light murmurations of knots and the small groups of wistfully trilling curlews. The seasons are marked by the changing avian populations: in the summer, ringed plover scurry on the shingle of the upper shore, the females wing-dragging to draw attention away from their nests, and there are larks and stonechats on the dunes. There is, too, the seasonal succession of flowering plants on the saltmarshes, dunes and shingle, and insects like dragonflies, butterflies and the striped cinnabar moth caterpillars on the ragweed. Even along the tidelines that shift up and down the shore through neaps and springs, recording the moon's passage around the Earth, the natural flotsam varies.

One of my favourite tideline treasures is a sandal, its sole decorated with the black leathery stalks and china-white plates of goose barnacles, *Lepas anatifera*. It was lost or discarded on the other side of the Atlantic, and the barnacles' planktonic larvae had bumped against it and settled. Growing by filter-feeding on other plankton in the sea, they had changed shape several times, metamorphosing and moulting their old exoskeletons. Their floating home had crossed the ocean, entered the Irish Sea and finally been cast up in the Solway. Goose barnacles, attached to flotsam as varied as sheets of polystyrene or old rope, have been arriving increasingly frequently in the autumn and winter, their arrival almost coinciding with that of the barnacle geese – but these days we know that is a metamorphic step too far. In late summer, drifts of dirty-white horn wrack are tangled among the débris, each seaweed-like 'frond' a flat colony of tiny 'moss animals', the filter-feeding Bryozoa. Or there are fragments of pale sponges, or the delicate skeletons of heart urchins, *Echinocardium cordatum*. A few kicks at the débris cause sandhopper hysteria as the little crustaceans leap to find new cover. Sometimes there are dead sea birds tangled in the wrack, their feathers dull with sand. A cormorant skull is still patched with skin, so I take it home and bury it in a net bag in the compost heap; after a few months' cleaning by bacteria, worms and insect larvae, its detailed architecture will be uncovered.

The tidelines also record the history of our own activities over days and years with a plethora of rubbish: paper, plastic, clothes, fishboxes and polypropylene ropes, toys and tyres. Its provenance and deposition depend on the wind and currents. Unfortunately most of the rubbish that enters the Solway piles up in various hot-spots on the northern shore such as Mersehead and Barlocco Bay, often in places where removal is difficult – but the Skywatch pilots of private aircraft, boat-owners and farmers have been enlisted to help.[10]

One July day, my husband John and I walk along the dramatic rocky coastline at Barlocco Bay and Rascarrel, sometimes on top of the high cliffs and above the caves and rocky pinnacles, sometimes on the shore. The cliffs are banded with colour – the pale, thin grass gives way to the bright yellow-orange of *Xanthoria* lichen, then the oil-black band of *Verrucaria* and finally the sea-scoured grey of the rocks. We drop down to the shore at Barlocco, and here the colours are blue, yellow, pink – a tangle of polypropylene and plastic amongst the wrack and branches. There is, too, a black cylinder, about thirty centimetres high and tapering slightly, a lip at each end: the entrance tunnel of a lobster-pot. The protective calcareous tubes that serpulid worms have built around themselves decorate the inside like white hieroglyphs. John knots a rope around the cylinder and slings it over his shoulder. At home he removes the base at the narrower end, replaces it with a circle of perspex, and adds two handles – a perfect bathyscope for peering into the water on my low-tide shore-walks.

*

I was out on the water in the fisheries' protection vessel, the *Solway Protector*. It was 2010, and at that time the English inshore fisheries were overseen by local Sea Fisheries Committees.[11,12] David Dobson, who was then Chief Fisheries Officer of Cumbria SFC, had invited me to join a patrol. The wind was a fairly gentle force 3 but it was immediately obvious what the phrase 'short seas' meant: because the Solway is a relatively confined space, the sea has no room to build up waves that have a long period between them, and this made for a bumpy ride. At fifteen knots[13] we left a widening wake of churned green water, but at about nine miles out we slowed down in parallel with a rusty Irish double trawler which had its two nets out and was motoring slowly north-west in a straight line; the CSFC officers waved and the skipper waved back. One of the officers, Alan, chatted on the radio to a Whitehaven trawler that was approaching us, the skipper grumbling that he had only 'five baskets of bulk'. Alan grinned at me and covered the speaker as the skipper swore, explaining that the trawler had caught nothing but 'rubbish' – 'junk, garbage, starfish and a few prawns, it hardly covers the diesel'.

The weather was starting to deteriorate and few boats were out fishing, so we turned and ran back along the coast on a following sea. Behind us the dark bulk of the Isle of Man was wiped out by a white curtain of rain, but the complex shapes of Sellafield's nuclear installations briefly caught the sun. We had been lucky with the weather; the Solway had not been unkind. But if a south-westerly gale coincides with a high tide, the effects can be dramatic. Seacroft Farm is perched beside the coast road at Dubmill Point on Allonby Bay; a friend of mine remembers his grandfather working in fields between the road and the sea, but the promontory has been subjected to such battering by stormy seas that it is now edged by a hard slope of concrete. The fields have vanished into the Firth and are probably now part of Rockcliffe Marsh. When a storm is due, the farmer almost literally 'battens down the hatches' and covers every seaward-facing window with external shutters. At high water the wind drives the waves against the hard defences, and pebbles and seaweed are hurled into the air and across the road, the spray from the breaking waves lifting high over the house. The drama always attracts photographers and voyeurs (including me). As David Dobson said to me, 'There's no two ways about it – the Solway's extremely dynamic!'

*

Norman Nicholson wrote of the Solway that 'even a stranger cannot deny its personality'.[14] So strong is that personality that the Firth is rarely referred to by that name – it is nearly always its first name that is used: the Solway. At the end of a long northern summer day the sun sets in the north-west, growing fat and red as it dips behind the shoulder of Criffel, splashing a palette of glowing-hot colours across the sky and sea. If you ask visitors and locals what they most like about the Solway Firth, the commonest answers are 'glorious sunsets', 'its beauty' and 'the peace and space': the benign and gentle characteristics of the greater land- and seascape.

But ask people who work on the Firth, and you only hear awed respect. To a haaf-netter it is 'one of the last wildernesses'; a plant manager of the offshore wind farm called it 'chaotic and unpredictable'; the development manager at the port of Workington said it was 'one of the most aggressive estuaries in the UK'. Marine biologists at Natural Power refer to it as 'highly dynamic', the Admiralty Chart marks areas of 'Changeable Depths' and a professor of coastal geomorphology said he didn't 'know of another estuary where so much sediment is available for distribution'. The skipper at the Silloth lifeboat station shook his head when asked: 'The environment is very, very unpredictable, it's uncontrollable.' (That final adjective is a warning for the Anthropocene.) And an artist told me, 'I feel there are loads of powerful, hidden things going on underneath the surface, even though the surface looks calm. There's an edginess, a tension that I love.'

'Hidden things' beneath the surface, especially the surfaces of the margins; the living creatures in the mud, the saltmarshes and the sand, which are rarely mentioned. These are the invertebrate animals, and although it is these uncharismatic yet ubiquitous invertebrates that populate the liminal areas where land and sea meet, their variety is scarcely known and remains unappreciated. David Attenborough's oft-quoted remark is so relevant here: 'No one will protect what they don't care about; and no one will care about what they have never experienced.' Or even seen. How many people living by a Solway shore where there are honeycomb-worm reefs even realise that the lumpy brown 'rocks' on the lower shore have been constructed, sand grain by sand grain, by small pink worms? How many people looking out at a saltmarsh fringed by mudflats – such a ubiquitous feature of the Solway and other major estuaries and firths – realise that the wet shorescape is home to millions of worms, snails and crustaceans which define, and are defined by, that edgeland. By considering them, our perspective is brought down from the satellites and gyroplanes to view the minute detail in higher magnification.

Many years ago I started to take people on low-tide walks on the Solway's Cumbrian shore. There are just a few walks each year, between April and September, because the best low spring tides are unsociably early and daylight is obviously a necessity. We walk straight down to the edge of the sea, preferably as the tide is still ebbing, and start to look at what is living in the sand, on the rocks and in the pools. If the water's surface is ruffled by the wind, we use the bathyscope to spy on burrowing sea anemones or sponges. Occasionally, depending on the time of year, a few small fish like young plaice or shannies might skitter across the bottom of a lagoon to hide in the sand or in crevices, but the other, multitudinous, creatures are always invertebrates. You have to 'get your eye in' – some people, irrespective of their age, are excellent 'finders' – and gradually the shapes and colours and behaviours of the shore's inhabitants become visible. The ever-observant Kathleen Jamie knows about that too: in *Sightlines* she advises, 'Keep looking, even when there's not much to see. That way your eye learns what's common, so when the uncommon appears, your eye will tell you.'[15]

The Victorian naturalist Philip Henry Gosse (1810–88) pioneered new ways of showing the wonders of marine creatures and their adaptations and behaviour to 'ordinary people'. He was the first person to lead shore-classes, and his son Edmund describes his father on the Devon shore:

I recall a long desultory line of persons on a beach of shells, – doubtless at Barricane. At the head of the procession, like Apollo conducting the Muses, my father strides ahead in an immense wide-awake, loose black coat and trousers, and fisherman's boots, with a collecting-basket in one hand, a staff or prod in the other. Then follow gentlemen of every age, all seeming spectacled and old to me, and many ladies in the balloon costume of 1855, with shawls falling in a point from between their shoulders to the edge of their flounced petticoats, each wearing a mushroom hat with streamers . . .[16]

This is how I like to remember Gosse when, wearing my wellies, salt- and mud-stained cagoule and unbecoming woolly hat, I lead my own line of chatting and not at all desultory 'persons' on my low-tide walks on the Solway shore. Gosse was an enthusiast for invertebrates, especially sea anemones (actiniae). He devised the marine aquarium (his own word) and, amongst the many books that he wrote, produced several on the collection and observation of the animals and marine algae (seaweeds) of the shore.

Indeed, the Victorians were so enthusiastically inspired to hunt for specimens for the glass-and-mahogany aquaria that graced their drawing-rooms, that Gosse later wrote to his son that, 'Years and years have passed since I saw any actiniae living in profusion; the ladies and dealers together have swept the whole coast within reach of [St Mary-church, Devon] as with a besom.'[17]

We don't collect living shore-animals or algae on my own guided walks, although a mussel shell with its layer of pearly nacre, or empty shells of the different colour morphs of periwinkles, *Littorina obtusata* – brown-and-yellow striped or orange or deep red – might be clutched in a hand or pocketed. The delicate white skeletons of heart urchins, *Echinocardium cordatum*, some with spines still attached, or balls of empty whelk-eggs that blow along the shore like tumbleweed, might also be taken home. We do collect, though, some ideas on how the burrowing or tube-building animals live, and how barnacles and sea anemones that are attached to rocks can catch their prey. It's fun to talk about how they arrived there, how they send out their offspring into the marine world, and how they behave when the tide comes in. But what we see on our walk is just a snapshot of their lives in their intertidal neighbourhood: we don't see the animals and the algae during that fourth dimension of time, when the water returns and rises above them in a shifting column of currents before gradually receding again, twice each day. But we might see the criss-crossing trails of now-hidden ragworms, the scrape-marks of a running crab's claws, or find sticky green balls of ragworm eggs stranded amongst the sand ripples, and these signs might help us a little to imagine what those busy creatures were doing after the tide came in.

*

Where I grew up, close to the south-east coast of Cornwall, the grey sandy beaches were enclosed by cliffs, and my father and I guddled for hours in rockpools and built dams on the shore, so when I arrived in Cumbria as an 'off-comer' twenty years ago, the wide flat beaches of the Upper Solway seemed quite alien, even a little forbidding. But our dogs, a springer and a border collie, both 'rescues', needed walking. The collie, a former sheepdog, remembered only the command 'Come by!' and she would run and run, clockwise in concatenated circles along the shore, leaving a trail that would surely puzzle a SkyWatch pilot. As we tramped the shores in all weathers and at all stages of the tidal cycle, I discovered the honeycomb-worm reefs, the mussel beds and lugworm 'nurseries'. The urge to find out more about the stories of the living organisms which inhabited this 'neighbourhood' – nature-writer Richard Mabey's perceptive term[18] – led me to ask 'experts', and of course I wanted to share what I had discovered, which led to the guided walks. But I was also meeting people who knew the Solway and its edgelands from the perspective of their work or family history, all of whom without exception were tolerant of this off-comer's questions and were very generous with their help.

For here, too, on both sides of the Firth, are the human stories of the people who live and work on the margins and on the water: the farmers whose stock graze the marshes; the stonemasons and quarriers of

New Red Sandstone; harbour masters and marine surveyors; the trawler-men and fisheries protection teams; artists and conservationists; geologists and engineers; aviators and coal-miners.

These might seem like two disparate threads to pull together, but the simple underlying theme of this book is how we all live together – for better, for worse – along the margins of this unique place. And because they are so much a part of my upbringing, education and academic research, I make no apology for occasionally giving the invertebrate inhabitants equal status with the humans. For we humans are a blip in time, albeit as a species with great and often careless power, and the Firth's margins have been home to thousands of plant, animal, algal and bacterial species – millions upon millions of individuals – since long before we arrived. They are the pioneers and they respond to the Solway's changes and, like the humans who came later, they also influence and change the edges themselves. They form the base-layer of the pyramid: the Solway's fish and birds rely on them; the Solway's fishermen and wildfowlers and birdwatchers need them too. Their range and abundance and diversity affect the 'neighbourhood' and the 'neighbours'.

I was squatting down by a brackish pool at the seaward edge of Calvo Marsh, sloshing water through a plastic kitchen sieve to separate muddy sediment from mudshrimps and – perhaps because of my euphoria at finding these little invertebrate animals on such an icy day – it seemed that these tiny *Corophium volutator* could be a thread in understanding the natural and human evolution of the Firth.

1

Invertebrates on the edges

Nothing is ever entirely as expected here at the edge of the sea. The ebb and flow of the tides can be accurately predicted and printed in the tide tables, based on calculations of the relative positions of the moon and sun and Earth. But the wind, the weather and atmospheric pressure have their subtle effects too: a band of high pressure holds down the high-tide level, or a strong south-westerly wind drives the waves up the Firth from the Irish Sea, or heavy rainfall on the Lake District Fells sends a bolus of sediment-laden fresh water rushing down-river to the estuary to be partially dammed by the rising tide. That place between the shifting tide-marks, that intertidal zone that changes its character twice each day from land to sea, can be a harsh place to live. As Rachel Carson writes in her beautiful and immersive book *The Edge of the Sea,* 'Only the most hardy and adaptable can survive in a region so mutable'.[1] One high tide is never the same as the next, especially in an estuarine environment like the Solway: the mix of saline and fresh water, the amount of sediment, the predators, parasites and planktonic food, will vary every time. And as the tide ebbs, fresh water gains supremacy in the central channel, while on the edges the plants, algae and animals are exposed variously to desiccatingly hot sun or wind, or rain, sleet or hail.

There have been periods when the cold was so intense that not only the fresh water overlying the saltmarshes and mudflats froze, so too did the edge of the sea. On Boxing Day 2010, small icebergs were thrown up on the cold white shore, some more than one metre wide and fifteen centimetres thick, and at the start of 2019 the sea in the Upper Solway was torpid with ice crystals and as sluggish as oil. But this was nothing compared to the winter of January 1881 when, during the neap tides, ice grew ever thicker along the Upper Solway's margins. At the next spring tide, the sea lifted the frozen plates and whirled them out on the ebb, so that they crashed against the cast-iron pillars of the railway viaduct that crossed the Firth between Bowness-on-Solway and Annan.[2]

> On Saturday night and early on Sunday morning, when the principal part of the damage was done, four men were on the bridge keeping watch . . . They could not see the ice through the darkness, but they heard it rattling and bumping against the pillars, and, hearing several times amidst the general noise, while the ebb tide was running between two and six o'clock, a sound which one of them compared to the report of a gun, they at once came to the conclusion that some of the pillars had been broken . . .[3]

The ice floes were reportedly as much as six feet thick, of all sizes; some of them 'suggested comparisons with fields that were one or two acres in extent'; some were as much as 100 feet long. On Tuesday, 1st February, a large section of the viaduct fell: 'the sound was tremendous, and the steel coming in violent contact with other portions of the ironwork threw off so much fire that the thick darkness was illuminated with a transient gleam of light'.[4]

To the onlookers who gathered on the shores throughout those days, the destruction would have been an exciting and awe-inspiring sight. Not so for the wildlife: the *Carlisle Journal* reported at the same time about 'A Hare in a Fix': 'On Friday a hare was seen on a block of ice floating down the Solway with the ebb tide. It was seen by workmen on the Solway viaduct, who were much interested in its efforts to get off, and out of danger. It was forced to stick to the ice, and soon it disappeared seawards.'

An image as iconic, although for the opposite reason, as a polar bear on an ice floe. Today, the photo taken on a mobile phone would surely have gone viral on social media.

It must have been a tough time, especially for all the small inhabitants of the saltmarshes and mudflats. Thirty years earlier, after a long spell of freezing weather in February 1855, 'the shores of South Devon were strewn with dead and dying Anemones . . . which rolled helplessly on the beach'.[5] Research into the survival of ice-bound marine organisms has shown that some, like mudshrimps, *Corophium volutator,* can actually survive in drifting blocks of ice, but nevertheless their death rate is higher and their populations decrease during normal winters.[6] Even without the extremes of weather, the intertidal mud and sand is a difficult place to make a home. There is nowhere to hide on these smooth surfaces, other than within them, or by constructing a dwelling upon them. Upon the sandy shores of the Solway, and especially in the Marine Conservation Zone at Allonby Bay, mason worms, *Lanice,* and honeycomb worms, *Sabellaria,* select sand grains and glue them together to make rigid tubes. Other polychaete worms like the lugworm *Arenicola,* sea urchins like the heart urchin *Echinocardium,* and bivalve molluscs such as tellins, *Macoma balthica,* shelter within the shore in burrows. Where the shores are muddy, burrowers include the clams and cockles, small ragworms, *Hediste,* and the mudshrimps. In all cases, the burrow needs one or more entrances so the animal can irrigate its home to get dissolved oxygen and food, and get rid of waste products and its eggs or larvae when the water returns. These animals, which only infrequently leave the safety of their dwellings, are completely reliant on the tides to give and take; a life in the intertidal zone means alternating twice-daily between relative quiescence – waiting – and frenetic activity. But although the mudshrimp *Corophium* finds safety in a burrow, it is rarely quiescent for it frequently crawls and forages on the surface even when the tide is out.

Our own lives are carried out on the surface of our world; few humans penetrate above or below that single plane, unless for work or in search of adventure. *Corophium,* though, has that extra freedom – to pass through that plane in either direction, from solid to liquid, liquid to air, at different times, dependent on the tide.

*

In the late 1960s I was a zoology undergraduate at Bedford College, which was in the heart of the artificially semi-rural setting of Regent's Park and at that time was still included in the University of London's collegiate system. Part of our curriculum was an in-depth study of the invertebrates, animals ranging through the evolutionary scale from the single-celled Protozoa up through increasing multicellular complexity via the Cnidaria, Annelida, Mollusca and Arthropoda, to the Echinodermata and finally the Hemichordata and other 'lesser Deuterostomia'. In those days, lectures and lengthy practical classes in the department's large teaching laboratory comprised a major part of each topic, and we worked our way through the animal Phyla, dissecting, and studying slides and live or preserved specimens under microscopes, drawing and taking notes. The demonstration areas would be laid out with tubes and jars, trays of prepared microscope slides, small aquaria and white enamel trays of water, a scattering of plastic forceps and glass petri dishes and rubber-bulbed pipettes. Formalin and alcohol fumes, the dank smell of lived-in seawater and the dusty caramel smell of dried exoskeletons, identified and classified those classes. The work was exacting and often infuriating. Dissecting an animal, looking for a particular, almost transparent, structure like a nerve leading to a simple 'brain' that was little more than a pinhead-sized ganglion, required patience and a steady hand, and a good deal of optimism too. But with perseverance and help, what beautiful underlying patterns and evolutionary adaptations were revealed. Such in-depth university courses, now disparagingly termed 'classical' zoology, that incorporated both the practical and the theoretical, went out of fashion several decades ago, partly because of the time and effort (for both lecturers and students) that they required. The ethics of the acquisition and purchase of specimens is also a valid consideration (although a saunter down a street in Hong Kong where 'sea food' – that is, marine animals – is on sale, might put the numbers that were collected for scientific study in perspective). But by looking, *really* looking, at the bodies of single-celled and invertebrate animals, and learning about where they lived, it was possible to appreciate the animals' importance and to see the relevance of subsequent studies on their physiology, behaviour and cellular and molecular interactions. It was a training in looking, and more importantly *seeing*, followed by thinking, questioning and attempting to understand. A training essential for a scientist – but necessary requirements for a fiction writer too. Even then, despite the many distractions of student life in London, I knew I was in love with the study of animals – not the larger furry ones, but those smaller, often hidden individuals, the 'animals without backbones'.

Pickled specimens, preserved in formalin or alcohol for later examination, are poor sad relics, having lost most of their colours and all of their flexibility. This is especially true of the various gloriously coloured sea anemones, such as the beadlet anemone *Actinia equina,* which is common around our shores, attached to rocks and pebbles in the intertidal zone. Its body and tentacles vary from rich, bright red to greenish-grey, and there are often bright blue spots at the base of the tentacles where there are concentrations of stinging cells, the nematocysts. Pickled and in a jar, the beadlet is a lumpen grey mass. The aquarium at Maryport has fine displays of living creatures from the Solway, including a shallow tank containing beadlet anemones which visitors can – very gently – touch. My grandchildren are thrilled when tentacles adhere to their fingers, and even more thrilled when they realise that the anemone treats those fingers as prey, attempting to capture them with a battery of sticky glues and poisons discharged from its stinging cells.

Philip Henry Gosse collected, observed and described sea anemones from all around the British coast – small parcels 'of a salt and oozy character' arrived with the morning post according to his son. In Gosse's *Actinologia Britannica* the origin of the specimens is carefully recorded and we read about Miss J.C. Gloag of Queensferry ('who has long been a successful *cultivator* of Anemones'), and David Robertson, who collected around Cumbrae and in the 'Frith of Clyde'. The antiquarian and naturalist Sir John Graham Dalyell sent anemones to Gosse from the Firth of Forth. He did not send 'Granny': she (it) was a beadlet anemone who survived for more than twenty years in his own aquarium. When Sir John died Granny was, according to some sources, given to the palaeontologist Charles Peach, who kept it for another twenty years. Peach had moved to Scotland in 1849, living first in Peterhead, then Wick, where he studied fossil fish: he too sent anemones to Gosse. Peach (1800–86) lived in Scotland at an exciting time: he was part of the scientific élite that included geologists John Horne, Hugh Miller and Sir Roderick Murchison, naturalist Edward Forbes, and the publisher and evolutionist Robert Chambers.

At low tide and in the region of the honeycomb-worm reefs in the Allonby Bay Marine Conservation Zone, we often find dahlia anemones, now called *Urticina felina*. They have a deep red column with red-and-white striped tentacles, but these details are often hard to spot because not only do the anemones burrow, they also try to hide themselves with a coating of shell fragments and pebbles. Seen against the sandy bottom of an intertidal lagoon, there is a pattern that is slightly, only very slightly, 'awry', but it betrays the anemone's presence. Gosse described and illustrated this species too (then known as *Tealia crassicornis*) – and it is one of the many exquisite glass models of invertebrates that were made by Rudolf and Leopold Blaschka, the Dresden glass-blowers.[7] Pickled anemones might be dull – but the Blaschkas' models of actiniae, most of which are based on Gosse's coloured engravings, glow with character and colour.

I still have my father's copy of Ralph Buchsbaum's 1938 book, *Animals without Backbones*, a slightly battered first edition, its blue cardboard cover decorated with a stylised drawing of a sea anemone and a very startled, cross-eyed flatworm that is rearing up on its tail. Buchsbaum begins his book: 'Anyone can tell the difference between a tree and a cow. The tree stands still and shows no sign of perceiving your presence or your hand upon its trunk. The cow moves about and appears to notice your approach.'[8] There is a delightful drawing of a cow grazing on a clump of grass beneath a tree. The line of his argument that leads from this apparently irrelevant introduction to the topic of invertebrate animals is amusing, taking in drawings of a coral skeleton decorating the mantelpiece of a very 1930s fire surround, and of a woman with crimped, permed hair looking down a microscope (and we should thank Buchsbaum for his recognition that women can be scientists too). We are implicitly, gently, chided for our ignorance: 'In this book we shall be concerned not only with the jellyfish, which is seldom seen by inland dwellers, but also with many animals without backbones, like clams, crayfish, earthworms and fleas, which are supposed to be already familiar to most people. In addition, many forms will be presented which generally pass unnoticed because they are too small to be seen without a microscope, because they live under water or in the ground, because they inhabit remote parts of the world, or simply because they escape the unobservant eye.'[9]

His line diagrams of animals are bold and easily understandable; I have a childish urge to colour them in with bright wax crayons.

In the Easter vacation of my second year at Bedford College, our compulsory field-course was based in the zoology department at the University of Swansea, where we studied marine and littoral biology for ten days. Each day we were taken by minibus to different shores: rocky, sandy or muddy – widely differing habitats – where animals and marine algae were adapted to live successfully in different ecological niches (the more scientific term for 'neighbourhoods'). We marked out transects, threw down quadrats and counted living organisms for quantitative surveys; we searched amongst red and green algae for polychaete ('many-bristle') worms, crustaceans and iridescent nudibranch sea slugs; with nets and spades and magnifying glasses we captured and examined invertebrates, trying to identify them from the damp pages of our *Collins Guide to the Sea Shore*. Back in the lab, animals and seaweeds were displayed in dishes and bowls of seawater, and the challenge of identifying each to species level would begin. It was taxonomy with the aid of dichotomous keys: if this, go to 2; if not this, go to 3

... tracking down through the alternatives of the minutiae of the natural world, burrowing down through the naming of parts to reach the final goal, the given name of the whole.

Some years ago, while researching my novel *Seaside Pleasures*,[10] I had several conversations with Dr David Brown, who was at that time the malacologist at London's Natural History Museum. He was the expert on African freshwater snails, having spent many years on that continent collecting, drawing and working out the taxonomic relationships of these snails. This was not for esoteric purposes but because so many snail species, like the bulinids, are of medical importance as the intermediate hosts (they 'host' the developing and multiplying larval stages) of the parasitic worm *Schistosoma* that causes the debilitating disease bilharzia in humans. Many of the snail species that David found had not been described previously and he told me – with great humility, for he was a kind and thoughtful man – that there are only a few people in the world who have that privilege of first description, for 'until a species has been described, it is unknown'. A species, or even an object, is not part of *human* knowledge until we are made aware of it. If he described, if he named, this animal, it became known to him, and through him, to others. I remember that he smiled when he said, 'It is as though I had created it.'

*

Allonby friends tell me that the local name for the honeycomb-worm reefs on the shore is 'coral' or 'sand coral'. (The subtidal reefs in the Firth are – possibly – called 'nar'! See Chapter 8.) *Sabellaria alveolata* is a polychaete worm, part of the major taxonomic group of Annelida that includes the lugworm, *Arenicola,* and the mason worm, *Lanice.* Its appearance isn't especially distinguished for a animal that re-engineers the shore – a marine biologist friend shrugged when asked what it looked like and said 'it's a worm' – but at one end of its three to four centimetre-long multi-segmented body is a mouth with two radiating crowns of pink tentacles.

Sabellaria arrives on the shore as a planktonic larva, a mere eighth of a millimetre long, transparent and delicate, swirled to and fro by the currents in the sea. There may be tens of thousands of its peer group in the sea around it, all needing to touch down on solid ground, and all at risk of being swept up and eaten by other filter-feeding invertebrates like mussels, sea anemones and even other sessile – 'fixed' – worms. The larvae are stimulated to settle where they 'smell' the cementing secretions of others of their kind. They settle on other *Sabellaria* tubes, piling in next to each other. Sometimes they construct a disorderly network of tubes, sometimes the tubes are straight as organ-pipes. It is the communal nature of the tube-building – the massed apartment blocks, the high-rises and the sprawling suburbs – that makes *Sabellaria* so extraordinary. Here on the Allonby shore they form small mounds and reefs; near Seascale further west they form platforms, and there are recently expanding platform-like colonies amongst the rocks at Parton and Fleswick Bay. The worms form these protective tubes around their bodies by glueing sand grains together with a sticky secretion. Dr Larissa Naylor, now based at the University of Glasgow, showed that the worms were fussy about the size and shape of grain they used as building-blocks: when she compared the particle size of tubes with that of the adjacent sand, she found that the worms tended to pick the coarser grains. They also preferred to use (as you might expect for an animal which has a soft, unprotected 'skin') grains that were flat and plate-like. This selection is accomplished without the help of eyes or hands – just by using the sensitivity of their tentacles. And tube-building is rapid: in the lab, Larissa found worms could build tubes up to five centimetres long within two months of settling, and rebuilding rates of four millimetres per day have been found elsewhere, where worm-tubes have been damaged by trawling.[11]

The Marine Conservation Zone at Allonby Bay (now part of the new Highly Protected Marine Area, see note 18, Chapter 9) was designated as such because of the importance of these reefs; despite the fact that many people, even some who live nearby, are unaware of them – and are often astonished when they see them up close. The shore here is a strange and mysterious landscape, where jagged sculptural forms are reflected in the still surface of intervening pools; in the greyscale light of early dawn, the reefs are black silhouettes puncturing the grey satin of the water. Wherever they live along the coast, the worms have changed it, their mounds trapping water, so that lower-shore animals and algae extend their territory upshore, where they can remain underwater for more of the tide's cycle. Now, on low-tide walks, we wade softly through these quiet pools, hunting for the burrowing sea anemones, shrimps, prawns, crabs, starfish and brittle stars, fish such as shannies, algae such as the sheets of *Lithothamnion* that paint the rocks pink, and the red Irish moss, *Ceramium;* here and there reefs are blanketed with the green breadcrumb sponge, *Halichondria.* Seen from a gyroplane, the pools and lagoons are like cells that glitter with light.

*

Of course, there are people who know about the invertebrates of the edges in a practical way, especially when the animals can themselves be eaten or used as bait to catch other animals. Someone told me how he used to hand-gather mussels with a rake and sieve; a photographer remembered his grandfather collecting winkles to sell; a friend showed me his traditional hand-nets for catching shrimps and prawns. One time I stood in the sunshine and listened as a man told me about fishing for sea bass while he dug for lugworms with a fork, teasing their fat red bodies from the muddy sand and plopping them into a shiny bucket, where they writhed and tangled. For several years groups of young men from the north-east coast of England have arrived on the South Solway shore with buckets and plastic boxes to gather unsustainably large quantities of soft 'peeler crabs' to sell as bait – they know when the shore crabs will be moulting and at their most vulnerable. Unsurprisingly the numbers have plummeted, yet two of the collectors grumbled to me in passing that 'there's no crabs left' on the shore.

One of the many enjoyable aspects of taking a group for a low-tide shore-walk is that other walkers, particularly if they have grown up on this coast, have so many stories to tell. Quite a few men (always men) tell us about setting out fixed lines of hooks on the lower shore, baited with chopped-up sea anemone or, better, sections of large ragworms. 'Some of the worms were three or four feet long – if you got one of those it would last you for days!'; they were species of *Nereis* (apparently now classified as *Alita*), brown rags, green rags: very active and predatory worms, paddle-like limbs each side of every segment, and with strong black chitinous jaws. I have yet to find one because they are active when the tide covers the shore, and hide under stones when it ebbs. But we find their footprints on the exposed sand and in the undisturbed bottoms of the lagoons – sinuously curving parallel lines, spaced two or even three centimetres apart, of tip-toe imprints. Down on the shore we see only a snapshot of life when the tide is out, but so much happens when we aren't looking; when the incoming water touches life, rises over it and the living organisms respond. Rachel Carson writes how 'the flow of time [on the shore] is marked less by the alternation of light and darkness than by the rhythm of the tides. The lives of its creatures are ruled by the presence or absence of water'.[12]

As well as identifying already-named creatures during that undergraduate field-course at Swansea, we were offered a range of short projects to help us learn about research: the constructing of hypotheses, the practicalities of 'materials and methods', the gathering and interpreting of data. Practical work, not theory, bringing the (shocking!) revelation that the lives and activities of intertidal animals worked to a different rhythm than our own, governed by the shifting clock of the diurnal tides and not the opening hours of the nearby pub. Evenings might have to be spent in the lab. I chose to investigate how the mudshrimp *Corophium* responded to the falling oxygen levels in its burrow – whether it became quiescent and sat out the low-tide period, or whether it pumped the water in its burrow more rapidly over its gills in an attempt to extract every molecule of oxygen. So how do you watch an animal that lives in a burrow? You fill a narrow glass chamber with some muddy sand from the vicinity of a *Corophium* colony, and hope the animals you have collected will construct their burrows next to the glass. You fill the chamber with seawater of different oxygen concentrations, you hope you can see the mudshrimp's delicate pleopods, those legs that carry the plate-like gills, and count their rate of beating . . . It was a short project, naïvely simple through ignorance, yet it held a mixture of logic, planning – and hope. I can't remember the results or my conclusions. All I remember is evenings in the lab amongst the gurgling aerators – and watching the rhythmic, synchronised beating of those pleopods, and the tiny animal's elegant, questing antennae. And that most of my friends were in the pub. I had no inkling that I would return to looking at mudshrimps in later life, celebrating the part they play in the life of the Firth.

*

Grey cumulus clouds are piled high in the pale wintry sky above the shining slopes of yellowish-brown mud that fringe the River Nith. The mud is overlaid by a couple of centimetres of sticky clay; our boots make sucking sounds with each step, and it's easier to move forward with a sliding motion. Adam Murphy leads us like the Queen's Guide for the Morecambe Bay crossing: slowly, probing with a long thumb-stick to test for 'quick-mud'. He and Andy Over are Nature Reserve Officers at the Caerlaverock Reserve,[13] and the three of us are on an 'expedition' in March 2019 to look for *Corophium*. There's no mystery about whether we will actually find the mudshrimps, but it gives us an excuse to walk out onto the wide, glistening bed of the Nith estuary, onto a transiently visible surface that is normally unvisited by humans.

Just beyond the saltmarsh's edge the mudflat is stippled with tiny holes and conical mounds of mud just a couple of millimetres high, betraying the openings of the mudshrimps' burrows. I dig the spade into the mud and lift it gently so that the divot of mud breaks apart but remains on the blade, and Andy – who is unfamiliar with mudshrimps – exclaims at the burrows that are exposed, some caught in section, others clearly U-shaped. In one of them, a small mudshrimp wriggles, its antennae waving. The burrows are barely two centimetres deep, and the shrimps seem small. Further down towards the falling tide there are drifts of empty shells of cockles and pink Baltic tellins, and here the *Corophium* burrows are densely packed and deeper, the mudshrimps larger, many at their full size of about one centimetre. The spade comes free of the mud with a loud slurping, sucking sound, water spurts from the burrow mouths, and mudshrimps seem to explode out, then crawl away. They are defined by their long antennae, nearly as long as their bodies. We watch a shrimp burrowing, and it appears to sink in, so rapid is its activity. Here the shrimps have burrowed deeper, some of their tubes reaching down five centimetres, and there are fine burrows containing slim brown polychaete worms, the ragworm *Hediste*, too. The mud on the spade is pale and below fifteen centimetres or so is interwoven with orange-brown fibres, possibly the decomposing roots of former plants. Andy collects some of the mud in my kitchen sieve, and swirls it in a shallow pool before tipping the captured objects into the white enamel pie-dish. Mudshrimps swim about, testing the meniscus and the bottom of the dish with their long, sensitive antennae. There is a constant flickering beneath them, the gill-bearing pleopods rapidly beating, the walking-legs scrabbling against the smooth sides of the dish. We all watch them, entranced; the mudshrimps seem so small and delicate – and yet, released, they can burrow, so very quickly.

It was perhaps too cold and too early in the year because we didn't see the other common inhabitants of mudflats, the tiny black snails, *Hydrobia*. A month later I visit Port Carlisle at low tide with Alison Critchlow, an impressively hardy artist who frequently works *en plein air*, and this time the surface of the mud is speckled with snails, hundreds per square metre. The initial impression is that they are immobile, withdrawn into their shells to wait out their exposure to the air, but the hazy sunlight flickers on their protruding heads as they feel and taste their environment, each moving almost imperceptibly on its broad, muscular foot. Scarcely two millimetres long, each snail ploughs a trail as it grazes, and the surface of the mud is scribbled with their journeys; Alison calls it 'the calligraphy of the shore', and later she sits on the edge of the saltmarsh to make watercolour sketches with quick, wide strokes of her broad brush, sweeping it across the page then pinching its bristles to a fine point.

There are other trails too, that she describes as 'juddering', for although the marks are only a little wider than the snails', many are patterned with regular cross-striations. Some of these trails are straight, some meandering; one heads determinedly in a line for more than half a metre then loops and spirals upon itself. Perhaps that long, straight line was made by a male, scenting the pheromones released by a female, then circling in to find her burrow. It is spring and the time of the full moon, when so many creatures release or enact signals that say 'I'm ready to procreate!' We squat down to watch as a pale-brown mud-shrimp crawls, 'judders', across the mud in the surface film of water, looping its abdomen underneath itself then straightening it: the striations on the trail are propulsion marks made by the fanned appendages of its 'tail', the uropods ('tail feet') and telson.

There is a hierarchy in taxonomy, where the greater is broken down into the smaller: for zoology A Level at school we learnt to chant, like a skipping-song, 'Kingdom, Phylum, Class, Order, Family, Genus, Species' – at that time blissfully unaware of Sub-Classes, Sub-Species, and many other levels and contentious groupings and ways of classifying living things, including much later groupings based on clades and DNA. *Corophium volutator* – Genus and Species; within the Class Crustacea; not in the Order Decapoda ('ten legs') like lobsters and crabs or the shrimps that come potted in butter, but in the Order Amphipoda; all of these Crustacea – Insecta too – within the Phylum Arthropoda: all animals with 'jointed legs'.

Just as taxonomists look for relationships between animals and, god-like, name the formerly unknown, so anatomists look for relationships and name the parts. Every part of the body has a name. Every segment of the jointed legs has a name. Even though I'm not an anatomist I love the music of the names, more skipping-songs. Watching the crawling *Corophium*, I struggle to recall the knowledge I had as a student, so long ago. With the edge of an empty mussel shell I try to draw the outline of a jointed leg on the mud's surface, to show Alison – the physical act of drawing helps retrieve the words: coxa, basis, ischium, merus, carpus, propodus and the pointy-fingered dactyl. I love the words for themselves, and their Greek or Latin derivations. I start to remember the names of limbs, too, and tell them to Alison, and she laughs: 'The naming of parts! It's fascinating. "Gnathopods" – I'm thrilled by even the *concept* of "jaw-feet"!' 'Pereiopods' – transporting feet. We don't need to know the names of the parts, but by knowing them and their provenance we can trace the beautiful economy of evolution – reuse this bit slightly differently, take a little bit of this and a little bit of that, try out the effect of this gene in controlling that later sequence and if it brings advantage, keep it.

Several days later Alison emails to say she has been sketching on the shore and 'a little creature crash-landed in my ink before thrusting itself off and across the paper . . . I looked more closely and it had the look of a mudshrimp . . . could that be the case? If so it might be the first instance of a mudshrimp-assisted drawing!'

*

In 2005 Stephen Burrowes, the Head of Science at Settlebeck School in Sedbergh, and I took a group of twelve- to fourteen-year-olds to the Solway shore at Allonby; they were of that intermediate age group – neither primary school children enthusiastic about 'nature', nor students studying a chosen subject for A Level – for whom science wasn't cool. Many of them had never been to the Solway coast, or any sea shore. We showed them a range of things on the shore in the hope that they would each discover at least one object that amused or interested them enough to find out more. The plan was that they would then write about it; whether as fiction, non-fiction or in a poem was up to them. They had all been told what to wear – warm clothes, no jeans, wellies not trainers. It was a cold, grey day but one girl, dressed in a short pink jacket and thin patterned leggings, told us she'd texted a friend to say 'I'm freezing my bum off here, but it's great!' I showed a group of boys how, if they paddled their feet up and down on the sand, it would 'liquefy' and they would sink in; stand still and the sand would then set around their feet, making it difficult for them to extract themselves. In other words, they were triggering a phase-change, an alteration of the relationship between particles and liquid, known by the splendid word 'thixotropy'. I took a photo that still makes me laugh, of three wildly grinning boys, their jeans (inevitably) stiff and dark with water, and their white trainers buried in the wet sand: thixotropy in action. Later, we all ate lunch in the warm and welcoming space of Jack's Surf Bar, watched by an oversized model of Elvis, microphone in hand.

Some students had found coloured bivalve shells, mermaids' purses, and the skeletons of heart urchins; they had found mussels and barnacles attached to rocks. Two girls tried to dig a lugworm from its burrow (but shrieking, ended up with a bisected worm; no doubt the worm was silently shrieking too). One of the boys wrote a poem about a lugworm (and, incidentally, thixotropy), which perfectly captures the day. To avoid embarrassment, I have changed the name of the lugworm-hunter:

> One day the lone lugworm was burrowing
> Through the sleek Solway sand,
> Then suddenly the sleek sand turned into
> A compacted mess.
> The lugworm was trapped.
> It finally wriggled free
> It was slithering faster and faster through the sand
> But the faint vibrations of the people could still be felt.
> The sound had stopped.
> But now he felt the sand being removed
> From above.
> It was Billy Smith digging away at the sand with his hands,
> He was only about 16cm away from the lugworm.
> The lugworm decided to wriggle to safety,
> The lugworm got caught in Billy's hands.

For a minute Billy was staring at the lugworm
Feeling its slimy skin.
He passed it to one of the girls.
She launched it into the air with disgust.
It landed on the beach unharmed,
Then it burrowed back into the sand to live another day.

Lugworms, burrowing bivalves like cockles and the pink-shelled tellin, and heart urchins – they all exploit thixotropy as they wriggle down into the littoral sediment to make their burrows. Thixotropy captures unwitting people and cattle in quicksands and mudflats too, which is why teams from Maryport's Coastguard and Nith Inshore Rescue practise rescuing people from mud as well as water.

More than fifty years ago, Dr Peter Meadows, one of my former colleagues at the University of Glasgow's Zoology Department, and his research student Alison Reid studied the burrowing and feeding behaviour of *Corophium*, which they collected from the 'fine grey mud' at Greenock on the Firth of Clyde.[14] They set up tanks and tubes in their lab, introduced individual mudshrimps, altered different parameters and, basically, used old-fashioned observation – their eyes and notebooks – to record what happened. So, when a mudshrimp burrows: '[it] apparently makes use of a thixotropic effect: agitation of the pleopods creates a furrow into which the body sinks . . . When the entire body is submerged the pleopods beat very rapidly, driving the animal further into the mud . . . The entire process usually takes 2 to 3 minutes.'

The animal sinks in, just like the ones that Andy, Adam and I watched on the Nith's cold shore.

*

On the western bank of the River Nith, and on the hottest day of the year, Peter Norman, the Biodiversity Officer for Dumfries & Galloway, and I glissaded through the slippery surface layer of the mud to look for the Nith's 'training wall' and rails (Chapter 4). Beneath our feet was the densest colony of mudshrimp burrows that I have ever seen; it was a marvel of utilised space and volume. *Corophium*, unlike other crustaceans such as barnacles and crabs, say, doesn't have a planktonic larval stage. When an adult female mudshrimp is receptive to a male, she releases 'come hither!' pheromones into the water, and a male comes swimming or crawling to her burrow. If they approve of each other, the male releases sperm which are swept into the female's special brood-pouch – her marsupium – where the now fertilised eggs develop into embryos. About two weeks later they hatch as miniature adults, ready to make their own way out, onto and into the local neighbourhood. In this way the *Corophium* colony builds up; as many as 10,000–50,000 burrows per square metre have been recorded. As a mudshrimp grows, so must it moult its old exoskeleton and expand the underlying new one, and on the Isle of Cumbrae in the Firth of Clyde, my friend and former colleague Professor Geoff Moore (himself an expert on amphipod crustaceans and *Corophium*) has seen drifts of cast cuticles 'lying along the tideline like snowflakes'. A crowded colony provides an easy meal for predators that enjoy the taste of mudshrimp. In Canada's Bay of Fundy, two to three million semi-palmated sandpipers descend upon the mudflats at the start of their autumn migration, and gorge themselves on the fat- and protein-rich mudshrimps.[15] Here on the Solway, predators such as waders take advantage of the low tide – I have watched little groups of redshanks picking off mudshrimps as they crawl in the surface film, the birds scurrying, snatching, trilling their satisfaction between beakfuls.

Judging by the webbed footprints and a few discarded white feathers, gulls had been resting and preening on the exposed training wall. Amongst the serrated wrack, *Fucus serratus*, their bronze-brown fronds dusted with a coating of grey mud, a few winkles, *Littorina littorea*, are hunkered down, feet and tentacles hidden inside their shells.

Winkle-collecting is a source of income along the Solway's Galloway shores, but is cold and back-breaking work, scooping up the slow herbivores as they graze on algae, or prising them out of sheltering cracks. How many winkles are needed to fill even one of the several large sacks that I see piled by the roadside at Borgue for collection? The winkles are sorted and graded by size – medium, large and jumbo snails – at a small factory in Kirkcudbright, and bagged up for export to mainland Europe. Boiled and pickled in vinegar and sometimes confusingly called willicks or even whelks, they are often sold at seaside stalls or in restaurants. In the interest of research I have eaten one or two – winkled them out of their shells with a wooden toothpick – and I am very happy to leave all winkles to their natural predators: herring gulls with their yellow-eyed supercilious stare, nervy oystercatchers and bossy, raucous black-headed gulls. Wherever they spend time resting on the rocks and scaurs is a good place to look for jumbo winkles.

Find a very large ambulating winkle, snatch it up and turn it over quickly to look at the large muscular foot before it is retracted into the shell and covered by the protective shield, the operculum. Does the foot have an orange tinge? If so, the winkle is parasitised by the larvae of a digenean trematode, a 'fluke'; the larvae are living in the snail's 'liver' or digestive gland, and the breakdown products of the liver cells include the orange pigment, a carotenoid. There is a marvellous interconnectedness here, between two species of invertebrate and two species of vertebrate on the shore, because that large winkle hosts part of the life cycle of a parasite, *Cryptocotyle lingua*, that is a common parasite of gulls, fish and winkles along our shores. In the probably unlikely event that you eat the uncooked winkle, the parasite larvae will not harm you – think of them as added protein, and stimulators of the more expensive jumbo grade of snail.

Instead, place the parasitised snail in a jar of seawater in the light and watch: soon it will be surrounded by a shimmering haze of *Cryptocotyle* larvae with long, forked tails, swimming towards the light. Hundreds, even thousands, of these cercariae might escape from the snail during a single day. They have two square eyespots and look as startled as, but so much more elegant than, Buchsbaum's flatworm. They are just one larval stage in a complicated life cycle that involves a snail (the winkle), a rockpool fish like a blenny or goby and finally, as an adult fluke, a sea bird like a gull. Living in the gull's gut, the adult flukes release hundreds of fertilised eggs which pass out in the gull's faeces and hatch. These new larvae swim around to find a winkle, then the tiny invaders burrow in, grow and multiply in the snail's digestive gland, eventually bursting out into the water as cercariae and swimming off to look for fish. A life cycle, a cycle of life. One egg, giving rise to hundreds of cercariae: multiplying to increase its odds in the parasite's game of chance, of finding not just one but three hosts.[16]

I love these parasites, for their elegance and for their extraordinarily complicated life cycles: imagining how this complexity arose through evolution is almost as difficult as was imagining the evolution of the mammalian eye before molecular biology uncovered the importance of certain controlling genes.

*

Some people collect and eat dog-whelks, too, though they look even less palatable than winkles. I have a favourite boulder on the Allonby shore, that my friend Ronnie Porter told me is called Hanging Stone – for its tilt, not its role (see Chapter 8). Coated in barnacles, it is a rock that always attracts *Nucella*, the dog-whelk, and in the spring each year dozens of them congregate here to mate. Suddenly, within two or three days – four to six tides – Hanging Stone and other smaller boulders are starry with the shells of the little whelks; they arrive in all their colours – dirty-white, grey, yellow, orange, banded brown-and-yellow – their shells ridged and aggressively pointed, unlike the drab and gentle winkles. They normally live scattered around the intertidal zone which, on the gently sloping Allonby shore, is often vast, and to get to the meeting points many of them must crawl great distances each time the tide comes in (they are generally motionless when the tide is out). The overhanging base of Hanging Stone is soon decorated with mats of their orange, vase-shaped egg-cases. Unlike winkles, they are predators: it's easy to anthropomorphise their behaviour as 'mean and sneaky' because predation by *Nucella* is persistent and patient, and their prey is entirely immobile – barnacles and mussels have no means of escape from the slow drill of the whelk's radula, its file-like strip of teeth, and the injection of digestive enzymes. Barnacles with empty, gaping skeletons, and mussel shells with a neatly bored, bevelled-edged hole, are signs of *Nucella*'s successful feeding.

Nucella was formerly known as *Purpurea*, because it could be used as a source of purple dye (the more famous Tyrean purple dye prized by the Phoenicians and Roman emperors came from the crushed shells of the Mediterranean clam, *Murex*). Gosse examined dog-whelks that he had killed, and found 'just behind the head, under the overlapping edge of the mantle, a thick vein of yellowish white hue, filled with a substance resembling cream: this is the dye in question . . . with a camel's-hair pencil you may paint upon linen or cotton cloth any lines, the initials of your name, for example'.

In sunshine the colour of the lines changed from yellow to green to blue, then 'the colour at length appears a full indigo', and eventually 'a dull, reddish-purple'.[17]

The web of connections between animals, algae and other living organisms on the intertidal shore is easily disturbed. Removal of large numbers of one species – peeler crabs, mussels or winkles – may have unintended consequences in their communities and neighbourhoods. The same of course is true for the subtidal invertebrates that we collect to eat. Until about ten years ago, one of the main catches discharged at Whitehaven's fish quay was 'prawns', the Norway lobster, *Nephrops;* living in burrows on the sea floor (see Chapter 7) they were caught by trawling, mainly in the Irish Sea, and over time the numbers caught gradually decreased – to such an extent that it is no longer economic to trawl for them. There are now Marine Conservation Zones in the Irish Sea where trawling is no longer permitted and where it is hoped the invertebrate communities will re-establish themselves. King scallops, *Pecten,* caught by dragging chain-metal dredges across the bottom of the Irish Sea, and the large common whelks, caught in pots, are the current invertebrates fished. I have no wish to eat a whelk for it cannot be

> . . . anything other
> Than a screw
> Of rubbery chew,
> A gurgle of goo . . .[18]

I visited Whitehaven's fish quay in early November and a scallop dredger registered in Peterhead, on the north-east coast of Scotland, had just arrived to offload scallops from the Irish Sea: the king scallop fishing season starts on 1 November. Moored next to it was a Fleetwood-registered boat: on the deck were about twenty red net bags stuffed with large edible crabs, and two sacks of lesser spotted dogfish – an 'uneconomic' by-catch – both species used as bait for catching whelks. The common whelk, *Buccinum undatum*, a large snail with a shell up to ten centimetres long, lives subtidally. Its empty white shells and the papery balls of its egg-masses that blow along the tideline like tumble-weed are common on the shore. Whelk are caught in baited pots that are strung along a line. A contact of mine in the Marine Management Organisation at Whitehaven told me that boats would occasionally bring in as many as 200 sacks of whelks, which would be taken to Fleetwood to be 'processed' then sent abroad, possibly to Korea, he suggested. 'What happens when their numbers decrease too?' I asked. 'Their numbers seem to be holding up . . .' he said. So far: invertebrates on the edge.

2

'Changeable depths'

We stood on the edge of the quay at Girvan and looked down into the boat that would take us out to the conical lump of rock that was Ailsa Craig, home to thousands of gannets. Small pieces of the engine lay on a wooden bench beside the open hatch, and the owner and his mate, cans in hand, stopped their conversation to look up at us. 'Aye, there's a wee problem. But we couldnae gae oot now anyroad, the wind's wrong. When it's in the north-east we cannae land, or if we did we'd nae get oot again.' There didn't seem to be much wind but the owner assured us with a broad smile that it would be blowing out there, and anyway the engine had a problem – and anyway it seemed he and his mate had things to discuss over cans of beer. So the five of us – the four geologists and me – turned away and prioritised Plan A: to go and look at rocks. I'd only agreed to accompany my husband because of Plan B's double reward of a boat trip and Ailsa Craig. Now I had to go and look at rocks.

But these rocks would be special because they were from the bottom of the Iapetus Ocean, the 1,000-kilometre-wide ocean that divided the ancient continent of Laurentia from Avalonia 450 million years ago. Steamy, subtropical Avalonia had broken away from Gondwana and had started sailing northwards, floating on the planet's molten core. And Laurentia, bulging north and south across the equator and containing a little piece that would become Scotland, was moving southwards. As the tectonic plates shifted they pushed against each other and, rather than crashing together, they came to a compromise and the edge of

Laurentia slipped beneath Avalonia. The Iapetus Ocean was gradually, very gradually, squeezed out of existence as the ocean bottom was subducted beneath the land masses, and proto-Scotland docked with the part of Avalonia that was proto-England, Wales and Ireland. By 300 million years ago, in the Late Carboniferous period, the 'United Kingdom' sat on the equator, a little speck within the massive continent of Pangaea until, still journeying north, it was released by tectonic movements and the rising oceans of Tethys and the Atlantic.

Our Plan A was to find the Iapetus Ocean: Iapetus, the god who fathered Atlas, after whom the Atlantic is named. Part of Iapetus's floor still exists, at Ballantrae near Girvan, and so we clambered around the rocky cliffs to find and stroke smooth lumps of pillow larva, and the 'ophiolite' layer that contains the hard, greenish stone, serpentinite. Ophiolites and serpentinites: sinuous, reptilian names – and the latter rock familiar to me from my childhood in Cornwall, where serpentine from The Lizard was polished and sold as ashtrays and paperweights. We were finding hints of the Iapetus Ocean in south-west Scotland, but the suture-line where it was finally stitched into the bowels of the Earth is in the Solway Firth. That 400-million-year-old memory lies close to the middle of the Solway, a line running north-east to Carlisle and across to Berwick. It may be old and deep, but perhaps it sends subterranean messages to the surface, where superficial weaknesses are exploited by sea and ice and river.

And so, skipping forward to the almost-present, a mere 12,000 years ago, geologists, geomorphologists, palaeobiologists and palynologists can explore and tease out evidence of the effects of water in all its states on the later development of the Solway Firth. But many of the clues are there for everybody to see:

– if you wander south along the shingle of St Bees beach on the Cumbrian coast, you soon reach a line of low, undulating cliffs. They are not formed of hard rock but of a jumble of rounded stones embedded in compressed red silty mud: pebbles, even small boulders, of New Red – sparkling with mica – and purplish Coal Measures Sandstone predominate, mixed with smoothed pale grey, greenish and black fragments of other rocks, scraped and swept and delivered here from the interior of the Scottish and English lands each side of the Firth. These cliffs are moraines, the crumbling remnants of the piles of rocky debris that were dragged here and deposited by the glaciers that covered this land. Friable, unstable, the edges will barely support a person's weight, and each stormy spring tide nibbles away at the evidence;

– further to the north-east, pebble-strewn hummocks, the remnants of glacial boulder-clay ridges, are seen reaching out into the Firth at low water – the rocky scaurs like Popple, Metalstones, Matta, Herd Hill;

– between Allonby Bay and Beckfoot, high tides and storms have sliced the seaward edges of the sand dunes, their newly cut vertical faces giving us a perfect view of conversations between the sea and land carried out over more than 10,000 years – layers of sea-washed pebbles, interspersed with sand and darker organic matter, each pebbly layer the evidence of a separate beach, laid down when the mean sea level was higher than today;

– both north and south of the inner Firth, on Kirkconnell Merse by the River Nith and on the intertidal shores of Beckfoot and St Bees, banks of rubbery peat and grey-green clay appear and disappear beneath the labile sand and mud.

*

'Once upon a time, woolly mammoths roamed across Scotland.' That would have been before the last Ice Age, about 30,000 years ago. In 2018 my friend Nic Coombey, who works for the cross-border Solway Firth Partnership, was walking on the shore at Loch Ryan near Stranraer when he came across an enormous bone. About half a metre long, it has been confirmed by the National Museum of Scotland as a femur from a mammoth that was probably about two to two and a half metres high. Unfortunately, not enough DNA could be obtained to help identify the actual species, but a sample has been sent off for radiocarbon dating. The bone might have been frozen into the ice or buried in clay, before being washed out and deposited on a sea shore many millennia later. For, approximately 20,000 years Before Present (BP), there was no Loch Ryan and no Firth – indeed there was no Irish Sea, for the whole area was smothered in ice which flowed southwards as far as the Isles of Scilly. So much of the planet's water was trapped as ice that the mean level of the sea beyond the icy margins lay as much as 130 metres (more than 400 feet) below its present level. When the Earth's climate warmed the ice melted, releasing churning torrents of fresh water that flowed down into the oceans. The sea level rose globally, spilling into the valleys and plains, making islands of hills, and pushing the coastal margins ever further inland.

But glaciers are weighty masses of ice and rock and, as they melt, the land is released and it 'rebounds', rising out of the sea. Professor David Smith, Visiting Professor at Oxford University's Centre for the Environment, and his colleagues from the University of Coventry, have extensively researched sea levels in the Solway as part of a European study into the effects of sea-level rise. David told me that 'The subsequent story is one of competition between rising sea and rising land, with both slowing in rate towards the present – the changes become progressively more subtle.' This subtlety and complexity has characterised the development of the Solway Firth, as David and his colleagues found during their fieldwork along the Scottish coast. Large deltas from the retreating ice are present in the Nith and Lochar Water valleys, with marine terraces along their seaward margins. Such terraces can be found throughout the Scottish Solway coastline, at heights of twenty to twenty-five metres above Ordnance Datum.[1] Then, the Firth would have been very much wider, reaching far inland to meet the torrents of meltwater that would have raced down from the ice on all sides. Anyone who has tried to wade across a meltwater river in Iceland will know how grey and gritty and almost unbearably cold this is. It would, then, have been an inhospitable place, and an inhospitably cold climate: there would have been animals and plants that were able to survive, but no Mesolithic human settlers would have ventured so far north.

Land uplift now increased and outpaced the rising seas so that the Solway would have become narrower and more shallow. David says its extent at this time is unknown, though he thinks it would have remained a major estuary. But then, by 12,900 years BP, the Earth's climate cooled again and the rise in sea level slowed or stopped as small ice caps developed during the Loch Lomond Readvance. Glaciers developed in the Galloway Forest Park area and in the Lake District. But the dance continued: renewed warming of the climate caused these new glaciers to melt and sea level rose, initially outpacing the isostatic rebound of land. However, the land eventually gained supremacy, and the old shorelines were left 'high and dry' – so we see the gravelly lines of raised beaches in the Allonby dunes, and the marine terraces and 'fossil barriers' of pebbles at different heights along the edges of the estuaries of the rivers Nith and Cree.

The impression given so far is of pebbly, gritty margins, but of course where rock is pulverised, sand and clay and mud are formed – and it is this sediment, borne by the sea and the rivers, that is so much a part of the Upper Solway's more recent story. For whenever the sea lost the competition, more land would have been exposed each side of the Solway. The estuaries would have been confined and narrowed, and become fringed by mudflats. Saltmarshes grew outwards onto the mud; dense patches of pale sedges whispered wherever fresh water was trapped, and salt-tolerant shrubs, then trees, colonised the landward edge: cold-climate vegetation at first, but succeeded by alder and willow carr, birch and oak. In this period, from 11700 to 7000 BP, trees grew. And it was a wet climate too, giving the peat-forming *Sphagnum* mosses a big advantage, so that bogs grew across the inner margins.

*

Saltmarshes classically develop in tiers as pioneer plants claim the mudflats and sediment accumulates on the surface; we can even watch this happening and – like Guillaume Goodwin and his doctoral research on Campfield saltmarsh (Chapter 4) – collect data and take measurements over several years; we might even set up experiments in the field or in a laboratory tank. But how does a scientist measure and understand the changing influence of the sea on a landscape that has developed over thousands of years? We can't do an experiment to test our hypotheses, we must instead find ways of examining the landscape in both gross and minute detail and then try to piece together the likely events. So geomorphologists like David Smith use surveying instruments – or more recently a GPS – to measure heights and distances and map the terrain; they collect samples of rocks and sediment to measure size and origin; they use gouges, power- or hand-driven, to obtain cores to help them analyse the stratigraphy, the different layers; they use light microscopes and scanning electron microscopes to look at pollen grains and the hard siliceous skeletons ('tests') of microscopic diatoms, to identify the species; other samples are sent away for radio-carbon dating to determine their age. It is a lengthy process, from the highly detailed and highly competitive applications for funding, to the laborious work, both out in the field (in changeable weather and amongst midges and biting flies) and in the lab, over a period of, often, several years. The results come from collaboration between many people, often of different nationalities, all with different ideas, expertise and personalities, and yet with the willingness to argue, bounce ideas off each other, assent or compromise, in order to reach some sort of answer to the initial question.

Here on the Solway Firth there are so many questions to ask, and it's easy to get frustrated because the answers are not there: perhaps no one has asked that particular question, and certainly no one has been awarded the funding to find out. In regard to his own research into changing sea levels along the Firth, David Smith has said, 'I am struck by the lack of evidence on the English side – research there has generally been very patchy', and a friend of mine who runs a Scottish environmental research institute told me, 'The Solway is the most under-researched estuary in the UK.'

*

Peat is here, all around the inner Solway – some of it hidden beneath the sand or mud and soil, some of it on show in the great raised mires, or Mosses, where the colours of *Sphagnum* mosses glow. Core samples gouged from along the banks of the River Nith showed bands of peat overlain by greyish silty mud, overlain in turn by another band of peat mixed with clay. The lower band of peat is 7,000–8,000 years old, laid down in the wet 'Atlantic period' of the Holocene. And this is where analysis and identification of diatoms is important, for some of these microscopic single-celled creatures, each enclosed in a highly orna-mented test, are characteristic of marine or brackish environments, others characteristic of fresh water (Chapter 5). In many cases the diatoms in the Nith banks' silty mud layers are typically marine and – although the data from several different research groups show the picture is very much more complicated than described here – this tells us that estuarine conditions occurred at least a couple of times, the incoming tides of the heightened sea level swilling sediment over the peatlands. Pollen analysis suggests alder and birch carr would have fringed the peat.

For several years, my low-tide guided walks included the shore near Beckfoot. We walked down the potholed track to cross the dunes, where marram grass whipped at our legs and thrushes had left 'anvils' surrounded by the shattered colour-banded shells of the dune-loving snail *Cepaea*. In the summer the stems of the yolk-yellow ragweed are decorated with the yellow-and-black caterpillars of the cinnabar moth, and spiky silver leaves accentuate the soft blue flowers of sea holly. Stonechats, red-breasted and black-headed, teeter boldly on sprays of *Rosa rugosa*, and the liquid songs of larks help pinpoint the singer high above. A short sandy path leads onto the beach and standing there on the pebbles, looking across the waters of the Firth towards the granite hulk of Criffel, you can see squat plates of darkness on the nearby shore – the patches of black are like shadows among the sharp-edged channels in the sand. As you walk down towards them the shadows resolve themselves into banks of something soft and organic. Poke them with your toe and feel how dense and sodden they are, smoothed by the friction of waves and sand. Walk on them and feel their sponginess, bend down and press the surface so that water oozes out, and you realise that the dark mass is peat. Your eyes gradually accept the patterns of difference around you – embedded here is a horizontal tree trunk, or there a stump that radiates roots. Wander around and you will find trunks and branches, single and entangled, and erect stumps whose tops are flattened as if severed by a chainsaw. The wood is still fibrous, soft and dark; you can crumble it and tease it apart, as though it were any rotting log in a woodland. But the difference is that this woodland thrived about 9,000 years ago. Take a small piece home and as it dries it falls apart into dust: like the wreck of Henry VIII's *Mary Rose* when it was first recovered, it must be kept wet to retain its form.

However, the Solway's tides and sands are capricious and the last time I saw this ancient submerged forest at Mawbray was in 2016; the trees were washed away during the subsequent winter storms, and by the following winter the peat too had vanished. Sometimes a section of the forest appears in Allonby Bay, or there is a brief sighting at St Bees. St Andrews archaeologists and the Solway Coastwise project recently found a peat-shelf on the banks of the River Esk by Redkirk Point with 'stumps, trunks and root systems of a prehistoric forest, identified as oak [jutting] out from this submerged former land surface'.[2] This evidence of the Solway's former edges is being gnawed and battered by the river. One person I spoke to, who regularly surveys the Upper Solway's shores, had never seen or even heard of the forest, and Brian Blake, whose very readable book *The Solway Firth* (1955) is illustrated by black-and-white photographs taken by J. Allen Cash, notes that 'Mr Cash went to Beckfoot . . . the submerged forest was not visible and I regret to say the residents he inquired from had not even heard of it'.[3]

Some years before he retired, I went to meet Brian Irving at the Solway Discovery Centre in Silloth to find out more about the forest. The Centre is in the converted red sandstone school, and the offices of the Solway Coast Area of Outstanding Natural Beauty (AONB) are at one end of the building, in the former Headmaster's House. Brian – bearded, bespectacled and slightly breathless with enthusiasm – was at that time the manager of the Solway Coast AONB, and he remains almost evangelical in his enthusiasm for everything to do with the Solway. Our discussion ranged from topic to topic, with frequent interruptions to ferret through books, and to check the maps of the coast and the channels and the other nature reserves that were pinned to the walls. What went on here was a reflection of everything else that was going on at that time in Britain, Brian told me. As the climate became warmer, 'there was a wave of plant colonisation spreading north, a "tundra front", with dwarf plants – dwarf willow and dwarf birch. The standard ecological colonisation.' Tundra vegetation: I have walked on it, sat amongst it, in eastern Greenland; overwhelmed by the silence, the scents, and the colours and shapes – juniper, and low, flattened willow plastered with fluffy seeds, dwarf birch and blaeberry blazing red and orange in the autumn, clinging to the sparse soil that is only inches above the permafrost. Imagine that carpet spreading across the land from the Mersey up to Silloth and across Dumfries and Galloway.

When, perhaps around 8000–9000 BP, the air and soil started to grow warmer, wind-blown tree seeds were able to survive and germinate. According to Brian, a pine forest gradually spread across our landscape, obliterating the tundra vegetation and lasting for a couple of thousand years. Mammoths, bison, giant deer and sabre-toothed cats grazed or hunted amongst the trees and on the grassland. The pines were succeeded by deciduous trees like alder, willow, birch and oak. This new forest was low and dense with a thick understorey; as Brian said, 'You'd have found it almost impossible to walk through.' But perhaps you would not have wanted to in any case, because this was a time when wild pigs were common. The forest would have been full of birdsong, and you might have caught sight of elk or deer grazing, or even the now-extinct aurochs, *Bos primigenus*, a large 'wild cow'. A friend told me that, as a lad back in the 1970s, he was fishing at Skinburness when he saw a 'row of pointed things' sticking out of a peat lens that had been uncovered on the shore. He and his friend went to investigate and, with the help of their bait knife, excavated the skull and antlers of a large animal like a deer. His memory is that they tied it to one of their bikes and eventually took it to a museum where it was identified as coming from an elk; sadly, Tullie House Museum in nearby Carlisle has no record of it.

It is difficult to put an age on the forest because there was so much variation in the sea level relative to the land, not only from place to place, but from time to time. David Smith thinks the time when the forest flourished could date from anything between 11700 BP to about 7000 BP, according to these fluctuations. The earliest submerged forest so far found on the Scottish side, in the Cree estuary, developed as early as 9500 BP but was buried beneath sediment within five hundred years. 'As you move southward,' David told me, 'and the forests developed later, the cover of later sediments becomes less and they are exposed at low tide.' He mentioned the submerged forests on the Lancashire coast at Formby, where footprints of aurochs and humans of perhaps early Neolithic age, around 7000 BP, have been found.

A section of forest and banks of peat emerged at Haverigg on Morecambe Bay early in 2019, not far north from where those early humans had left their footprints in the red clay. The tides dispersed it quite quickly but not before it was noticed that one block of peat was pitted with holes, many of which contained the shells of piddocks. Knowing that I too had found piddocked peat at Beckfoot, someone sent me a photo – and, as with the Beckfoot piece, I was yet again thrilled to see the remains of these 'boring bivalves'. Piddocks, *Pholas*, are molluscs that, like mussels, are enclosed in two elongated shells, in this case decorated with bands of jagged serrations. The animals live below the intertidal zone, normally tunnelling their way into rock (peat must have been a delightfully easy option) by shoogling their file-like shells. Feeding by filtering food from the water that passes across their gills, they enlarge their burrows – but not the entrance – as they grow, so they remain locked *in situ*. For these molluscs to have grown to several centimetres long tells us that this peat must have been uncovered on the seabed for several years before disappearing beneath the sands again. It is indeed possible that a 'peat horizon' still lies across the bottom of the Firth and is intermittently exposed. Several years ago, Silloth's then harbour master and I became very excited because the latest bathymetric survey of the channel between Workington and Silloth had shown a large flat area between the shipping channel and the Beckfoot shore. We became convinced it must be a layer of peat, and plotted how we could persuade someone to dive into the murky waters to have a look. Unfortunately, subsequent conversations with the hydrographic surveyor down at Barrow suggested we were wrong!

The death of that ancient forest was probably due to several reasons. In places, the rising sea oozed amongst the trees, depositing salty sediment and débris amongst the roots so that salt crystals blocked the water channels within the tree trunks and the branches died of dehydration. 'It would have been like beaver damage,' according to Brian Irving. 'The twigs then the branches die and fall off, and you're left with the spikes of stumps sticking up in the water.' Or the forested areas might have been drowned by rain; about 7,000 years ago, in the Atlantic period of the Holocene, the climate was especially wet, the hills were almost permanently hidden in cloud and the rain poured down, day after day. Plants struggled to survive in the waterlogged ground; trees died. The stark silver skeletons of birches that mark the line of the former Solway Junction Railway across the raised bog of Bowness Moss (Chapter 5) show the effects of drowning in recent times; their uncluttered leafless branches make perfect viewpoints for crows and raptors, whose arrival causes whistles of dismay amongst the rafts of wigeons that laze on the water. This new climate was also perfect for the growth of *Sphagnum* mosses. As the rain fell, the mosses soaked up water while most of the other vegetation died and decayed and became compressed beneath the proliferating *Sphagnum*; this layer of peat spread, blanketing the valley floors and creeping up the hillsides, burying the remains of the forest. Just a short distance inland, domes of peat were growing upward, ten centimetres every century, to form the raised mires. On the plain and around the edges of the Firth, the forest was gradually hidden and preserved in an anoxic, acidic peaty bed. Then yet again the sea level rose, and the forest and the peat were submerged. The sea crept in across the soggy land bringing its own biosphere, and land-based life retreated from its incursion.

*

At the edge of the dunes near Allonby erosion occasionally reveals thick bands of clay underlying the layers of pebbles and dark organic material, but this material is higher and younger than the clay on the shore. The latter, a grey-green clay that underlies the peat banks, is smooth and slippery, a fine-particulate outcome of glacial grinding. Some of those students on the Settlebeck school-trip (Chapter 1) who discovered thixotropy also discovered the delights of sliding across the clay's surface. One of the boys told me his brother was a potter so – rather sweetly – he wanted to take some clay home for him; ten minutes later his arms were plastered grey and his lunch-bag was sagging with the weight.

There is red clay too. The longshore drift along the Solway's southern coast carries sand and shingle eastwards, towards the head of the Firth. Wooden groynes jutting out into the sea from Dubmill Point and Silloth try to hold back the flow but struggle to assert themselves as pebbles are sorted from sand and pile up against their westward sides. And as the sand and shingle are shifted from the beach so, occasionally, are patches of red clay uncovered on the middle shore, perhaps derived from finely ground red sandstone, with a texture that is markedly different from the smooth and slippery clay beneath the peat. Shiny when wet, dulled and cracking when exposed to the sun and wind, the red layer is like a tablecloth on which are displayed the multi-coloured pebbles of the shore – grey Criffel and pink Ennerdale granite, dark Skiddaw slate, coal fragments, milky quartz and the blue and moss-green bubbled fragments of Workington slag. Its redness can be scooped up with a stick or knife, gathered and rolled and fashioned into a ball. I have such a ball, slightly pinched at one side and perforated by a hole, that I found amongst the shingle. One of my friends has found a couple of dozen over the years; like mine they are four to five centimetres in diameter, weighing between fifty and seventy grams, and in many of them the clay around the hole has been worn thin. Loom stones or fishing weights: they could be either. In the museum at Dumfries a partly woven cloth hangs from a stick, the warp threads weighted with similar red clay weights. My Allonby friend found several lying in a line on the lower shore where a rocky scaur had been newly exposed – it's easy to imagine them weighting the bottom of a gill-net. Whatever their uses, some reportedly date from Roman times: the Roman army had encampments and mile-fortlets stretching down the coast from the western end of Hadrian's Wall – enough people for there to be a need for a burial ground near the Bibra Fort near Beckfoot. Some of the spherical weights, listed on the British Museum's 'Finds' website, may be even older, from Neolithic times. Irrespective of their antiquity, they are comfortable artefacts to hold, and I like to imagine children from the forts scraping at the damp clay, squeezing its clammy roughness in their hands, patting it into balls, poking them with sticks to make a hole; their hands and faces smeared with terracotta streaks.

Clay has been an important building material on the Solway Plain, too, since at least the fifteenth century. In a landscape formed on glacial till, gravel and mud, with very little native rock, how do you build a dwelling? You use the materials to hand – the earth and clay and straw, and whatever trees you can fell for timbers. Stone is needed for a plinth as foundation for the walls, lest rising damp gradually liquefy the clay construction; perhaps the ruins of an abbey or the Roman wall can provide a source, otherwise cobbles or field-stones must do. And if friends and neighbours help to tread and mix the clay with straw, then the walls will rise quite quickly. 'It's conceivable that it could be done in a day, using lots of people from the village, say. People with the fitness of the eighteenth century, rather than someone like me who stops for an egg sandwich!' says Alex Gibbons, a William Morris Craft Fellow who specialises in earth buildings and was involved in building the 'demonstration' clay dabbin house at Campfield. The advantage of the dabbin method is that it is quick.

Peter Messenger is a local expert on the Solway's dabbins, and has written an informative illustrated article with practical instructions about their repair. The mixture uses different ratios of clay according to its availability locally, but it 'could contain 30% (by weight) of stone/ gravel (from 5mm to 40mm); 30% of coarse and fine sand; 15% silt and 25% clay. There are examples on the Solway Plain where the proportion of silt and clay in total can be as high as 80% and these walls are as hard and compact as others which have 50% of stone and gravel. So there are no hard and fast rules'.[4]

The whole becomes, essentially, a composite material when straw is added – the straw lends tensile strength and prevents cracking. The amount of water added to the mix is critical. The well-trodden mix is then lifted onto the wetted plinth, and spread and trodden again before another thin layer of straw is overlaid. These interleaved layers of straw act to suck out the moisture from the mixture, and because all the layers are thin the wall can be built to its full height without having to allow intermittent periods for drying out. Lintels for doors and windows were put in place as the building rose; traditionally the supports for the roof were the inverted Vs of wooden 'crucks' tied together by horizontal cross-trees, often with purlins running the length of the roof. Various materials – including turf and heather – were used for thatch, and the walls were coated inside and out with lime-render, to prevent rain penetrating the dabbin and causing it to slump.

An early twentieth-century survey found about 1,500 dabbins on both sides of the Border, but by the time Nina Jennings carried out her own survey nearly twenty years ago there were only about 300 remaining. Her book, *Clay Dabbins: Vernacular Buildings of the Solway Plain*, is the classic reference book,[5] containing entertaining stories of some of the home owners. Jennings herself was an extraordinary woman, who studied for a degree in electronic engineering and an MSc in electronics, was a member of the anti-war Committee of 100 (her active involvement in protests led to her brief incarceration in Holloway prison) and of the CND and, amongst other activities, a local historian and a keen walker and skier; she died in 2015.[6]

The dabbins on the Scottish side of the Firth and Border have all but disappeared but, according to Peter Messenger, much of the land in

Cumberland was held by 'customary tenure'; by the middle of the nineteenth century, these customary tenants had to be treated as though they were the landowners – as long as they paid their rents and fines, they could not be dispossessed of their property. This explains why many more clay dabbins remain here on the Solway Plain. But Peter's own surveys found that many of them were in a state of disrepair, with damaged rendering and unstable walls caused by water ingress. He, Alex Gibbons and Chris Spencer of the Solway Wetlands Landscape Partnership realised that to help people – owners, surveyors, builders – understand how to protect and repair this special type of vernacular building, a practical demonstration would be useful, as well as being an entertaining project that could gather in local volunteers of all ages. Thus in April 2016 the clay dabbin building project was started, with financial and other support from a large number of organisations, and was constructed on the RSPB's reserve at Campfield, near Bowness-on-Solway. Penrith red sandstone blocks were used for the foundations, and the reddish clay and soil mix for the dabbin came from about 500 metres away, dug out when one of the reserve's ponds near the edge of the Firth was extended. Hundreds of volunteers helped during that year – land agents, building inspectors and RSPB wardens, retired folk and conservation volunteers of all ages, trampling the mix, lifting it into place and, later, helping with the roof and floor. A photograph on the information boards inside the completed building shows a group of happy primary school children in hi-vis jackets and helmets moulding the clay with their hands and forks, their arms smeared red up to their elbows.

*

So when did humans arrive on the Solway? Perhaps people were exploring the margins even as the sea level rose and the wildlife-rich forest and the peatlands were overwhelmed. We think about changes in thousand-year blocks of time, but in human terms this might have equated to roughly thirty generations, during which period the relative heights of the sea and land continued to alter. People would have moved away and returned, and each time they came back, generation after generation, they would probably recount stories from their collective memories of how the sea had hidden the peatlands, had formed new beaches and built new marshes and mudflats. The changes along the shores would have been just perceptible within each single lifetime. These early settlers and wanderers would have worked with the tides, fished the rivers and estuaries, roamed the saltmarshes and shingle and the rocky shores. We know that by the time of the Bronze Age, 3,000 years ago, greater numbers of people were living and moving around both sides of the Solway, between the shores and rivers. Humans slowly began to make their mark along the Firth, at first working with the natural changes and later making changes themselves.

Writing about the origins of human settlement in what he calls the 'Solway country', Alan Scott says they 'are swathed in clouds of uncertainty'.[7] There is some evidence of settlement even about 9,000 years ago, but the picture is becoming clearer moving from the Neolithic to the Bronze Age, as archaeologists find human modifications of the landscape like enclosures, stone circles and habitations. The various museums in Dumfries, Annan and Carlisle have collections of artefacts like tools, weaponry and jewellery. Unsurprisingly, given the labile nature of the edges of the Firth, people tended to live further inland on the higher ground, but the Romans clearly perceived the importance of the coasts on both sides of the Firth in terms of defending their territory. They started constructing Hadrian's Wall in AD 122, between Wallsend in the east and, passing just north of Carlisle, to Bowness-on-Solway in the west. This western end of the Hadrian's Wall Route for cyclists and walkers is now marked by a pagoda-like structure containing fine mosaics depicting wading birds and the much-loved winter immigrants, barnacle geese. There is, too, a bench where the weary can sit and watch the changing tides engulfing and then revealing the waters of the Eden and the Esk. But the fortifications do not end there – the waters of the Solway remained a potential route for invasion. So mile-fortlets and towers marched at least twenty-seven miles down the Cumbrian coast, from Maia at Bowness past Bibra at Beckfoot and so on to the large and important fort of Alavana at Maryport. From these coastal defences the soldiers would have been able to see any attempts by marauders from the north to sneak around the end of the Wall, or sail up the Firth. Built of timber and turf, and supplemented in places with palisade fences, the fortifications have not survived well, but aerial photographs in 1968 revealed the marks of ditches at mile-fortlet 21 on top of Swarthy Hill on the coast at Crosscanonby; the fortlet was excavated in 1990 and subsequently partially reconstructed.

The aspect of the coast would have changed as settlements grew up near many of the forts and fortlets: perhaps people felt safer, and the garrisons would in any case have needed servicing, in a variety of ways. The soldiers came from many different nations across the Roman Empire: at the fortified church of St Michael's at Burgh-by-Sands (where the body of Edward I lay for a few days before being carried south), which is a few miles east of Bowness and on the site of the Wall fort of Aballava, there is now a plaque: 'The first recorded African community in Britain guarded a Roman fort on this site. 3rd century AD. A BBC History Project.' The unveiling of the plaque was filmed as part of the BBC series *A Black History of Britain,* presented by David Olusoga. Alavana at Maryport was garrisoned by Spanish soldiers. Maryport's Senhouse Museum is crammed with Roman artefacts, including piles of stone altars and engraved stones showing fine horses (and what appears to be, but cannot be, a woman toting an elegant handbag). After the Romans left in about AD 410, their Wall and other buildings became an excellent source of stone for local buildings; the builders of St Michael's Church at Bowness made good use of its position at the end of the Wall, and altar stones make unusual additions to the walls of a fortified farm at Burgh and a house in Port Carlisle.

Brian Blake's accounts of the waves of settlers and invaders that swept into the Solway area, by land and by sea, from the Romans onwards, are a delightful read, and Alan Scott writes of the Norsemen (Vikings) coming into the Firth from Ireland while the Danes travelled overland from the east. He links the various suffixes in the names of villages to the languages of the invaders, and we see from both authors' accounts how the ownership of the Solway region swept back and forth, from the Celtic kingdom of Rheged to the Scottish kingdom of Strathclyde, finally being divided between England and Scotland by a line on the map. Invaders and quarrelsome neighbours used the Solway and its crossings, the 'waths' (Chapter 8), as well as the non-existent or porous land border. How much they 'remade' the changeable margins of the Firth, if at all, is unclear. More obvious changes were probably made by the medieval monastic houses both sides of the Firth: clearance of trees from forested areas of the Royal Forest of Inglewood, which extended across the northern end of Cumberland, reclamation of land from the bogs and marshes, and improved agricultural practices. The Cistercian monks at Dundrennan and Sweetheart Abbeys on the Scottish side and Holme Cultram at Abbeytown on the English side were notable for their agricultural influences along the coasts.

Then there were the nineteenth-century industrial influences, and the more recent – though scattered – signs of the Solway being used as a destination for tourists: large caravan and trailer parks on both coasts, concrete promenades where one might 'take the air', watch the waves, and fish.

*

The major industries like coal mining and iron making have altered the coastal fringe and left their marks. To the north of Workington and not far from the port, a cliff is banded orange and purple and red, like a section through an old volcano. Low rusty-brown cliffs are bolstered by tumbled blocks of limestone, and reefs of slag spill towards the sea in solid waves, part hidden by slippery yellow-brown fronds of wracks, *Fucus vesiculosus* and *F. serratus*. Sea-smoothed pebbles, grey, green, blue, all sizes, and some as bubbly and light as pumice, have drifted over the hard pavements of accreted slag. It's a landscape that intrigues and draws you on, further and further along the beach: a man-made landscape, and you can only wonder at the human effort it required, the sequential steps – quarrying, haulage, smelting and yet more haulage.

West Cumberland was ideally suited to the production of iron and steel. Haematite and limestone were available to be quarried; and so was coal, both beneath the land and beneath the sea. In 1856 the Workington Haematite Iron Company Ltd, set up to make pig iron from locally mined haematite ore, had two blast furnaces at Oldside just north of the town; Bessemer steel making commenced there in 1877. In that same year the Workington Moss Bay plant added three Bessemer converters to its own ironworks, and the production of pig iron and its conversion to steel escalated within the county. Sadly, the story of the industry's decline is well known: Cumbria's iron and steel works closed in the twentieth century, while Corus's rail-making factory lasted until 2006. Nevertheless, the Workington area was a centre of iron and steel production for a hundred years, and during these decades, one major unwanted by-product was slag.

Iron ore, limestone, coke: to make iron, you need these three ingredients, plus heat. 'Coking', or metallurgical, coal – in contrast to 'thermal' coal which is used for generating power – is produced by heating suitable coal to very high temperatures in the absence of oxygen to drive off the impurities such as sulphur and phosphorus and create a hard porous material. The mixture is introduced at the top of the blast furnace and hot air is blasted into the furnace near the base. As the hot air burns the coke, the resulting carbon monoxide reduces the iron ore to iron. Molten iron is tapped off into a channel with lateral chambers to form iron 'pigs' (the pigs can then be converted to steel in a Bessemer converter). The limestone acts as a flux, combining chemically with coke ash and impurities from the ore to form slag, which floats on top of the molten iron. It was tapped off into specially designed trucks, called ladles, and taken away for disposal. This still-molten slag was tipped along the coast near Workington, forming artificial hills and cliffs, pavements and reefs. Local historian Russell Barnes describes the hazardous business of emptying each ladle: '[It] was held in place on the track by ramming a giant wooden wedge – called a "Scotch" – under the wheels. The loco then ran backwards, tensioning a chain attached to the ladle. As the chain tightened, the ladle tipped its load, returning under gravity when empty.'[8]

One of his photos shows white-hot slag cascading down the tip, and he remembers '. . . standing on the green metal steps of the old Rocket-Brigade station with my Granda – half frightened, half excited – watching this regular, awe-inspiring event. Several tons of white-hot slag racing downwards from above; Workington's own Vesuvius.'

And so the beaches are strewn with man-made artefacts of coloured and 'floating' stones, and rusty iron-bearing material that resembles the smooth and oozing shapes of solidified lava. There's a strange life-lessness to the shore, sparse in visible marine life except where winkles crawl amongst the algae that patchily coat the reefs.

When I visited it was also a place of alien sounds: the regular scything 'swish' of nearby wind-turbines, reminiscent of a samurai sword; a man shouting angrily above the baying and yapping of an intermingling pack of dogs along the shore; the growling of trail bikes unseen behind the slag banks. But at the top of the shore hemispherical mounds of sea-kale, with fleshy leaves and small white cruciform flowers, spread defiantly across the banked-up pebbles, and a short way inland, heath-spotted orchids and wild roses decorated the once-derelict land.

*

The water had filled Marshall Dock at Silloth and the shrimp boats were nodding gently on the ripples. You can see right across the Firth from the windows of the Harbour Office, a broad – and on this day – flat, blue stretch of water, bounded to the north by Criffel and the Scottish coast. What you cannot see are the 'Changeable Depths'. Ed Deeley, at that time the harbour master at the port of Silloth, and a ship's pilot, had spread out the Admiralty Chart showing the approach to Silloth Docks; land was coloured yellow, the foreshore and sandbanks – Silloth Bank, Beckfoot Flats – pale green; the sea was, unsurprisingly, blue – and there were pale pink areas 'Unsurveyed'. Figures giving the depths were printed all over the watery areas, including large patches of 'Changeable Depths'.

'Between Solway Buoy and Corner Buoy, it's a critical region, the region that gives us the most trouble. At Corner Buoy there's a narrow corridor – that channel is our window to Silloth, to the east of it are big boulders, the shrimp fishermen won't go there; to the west are shifting sands.' Ed explained one of the major problems of bringing a ship up the Solway Firth – the sea constantly re-sculpts the seabed. Sediment is constantly on the move, being scooped up, suspended and deposited. This affects fishermen – trawlers, scallop boats and shrimp boats, lobster-pot men and haaf-netters – and vessels that service the wind farm, as well as the larger ships that transport goods around the coast. Ports and harbour entrances need to be dredged, and the depths of the channels surveyed and re-charted.

Pilots taking ships up to Silloth take note of the depths registered on the ship's sonar, especially after storms, and in recent years the route from Workington up to Silloth has been periodically surveyed by Associated British Ports Ltd (ABP), who own the ports of Silloth and Barrow on the Cumbrian coast.

Chris Heppenstall, ABP's hydrographical surveyor, was very willing to show me the data and talk about the surveying, so I went down to Barrow to meet him. The port office is a red sandstone Victorian building, now almost hidden from its *raison d'être* by a large office block built on an area where once there would have been warehouses and cranes. Barrow is still a large and important port and home to a ship- and submarine-building industry, and Chris explained that the deep, straight channel from the docks, out past Walney Island, needs survey-ing every few months and dredging once a year, sometimes more. But as well as surveying the approaches to his home port, Chris goes up the Solway to Silloth. The multibeam scanner on ABP's tug sends out a broad fan-shaped pulse of sound, which bounces back, and the length of the time delay creates a picture of the topography of the seabed. But the boat isn't stationary, it's rolling and moving up and down – so the scanner's position must be accurately determined by GPS and the signals that bounce back from the bottom have to be corrected for factors like wave height, and the pitch and yaw of the boat. 'There are swathes of coverage, we try to overlap the corridors,' Chris explained. 'The area scanned depends on water depth – it's greater when there's more water. Obviously, if you're vertically above a spot, the beam has a smaller footprint, so you get greater detail.' But the port of Silloth makes life difficult because it can only be left or entered during a short period each side of high water, and surveying is very weather-dependent. Recently, they have been based at Maryport and have used a new twin-hulled boat instead of the larger tug.

However, the ways in which the scanner results can be displayed is impressive. At Silloth I'd already seen large coloured print-outs of the charts, showing the position of the survey relative to the three buoys. Now, Chris showed me multi-coloured 3D images on his computer, the colours relating to the measured depth. Images from different surveys could be stacked and tilted, to show progressive changes in the height of the seabed. The detail was extraordinary – I could see the large, regular sandwaves in the north-west section of the survey and a scattering of boulders across one of the submerged scaurs. By comparing the surveys from different periods, we could see how the sandbanks had moved eastwards, partially filling the Silloth Channel, and then westwards again a few months later. The pattern of constant flux is still continuing: the latest surveys show that the sandbanks to the north and north-east of the channel move to and fro, but the central and western parts of the channel overlying boulders and clay remain fairly stable.

Despite this glimpse into the shape of the Firth's undersea, we know surprisingly little about its inhabitants. What do people normally envisage when they are asked about the undersea? In 2008, 99 per cent of 4,000 people surveyed by Natural England couldn't name a feature or creature associated with the general undersea landscape and 44 per cent thought it was 'barren'. More upsettingly, for 60 per cent of the sample the instinctive response to the undersea landscape was 'characterised by a mixture of fear, disgust and shame: fear because it is a dangerous place, disgust because it is thought to be cold, dark and slimy . . . and shame because it is thought to have been allowed to get into this state . . .'[9] Back in 2003 I walked out across the 'undersea' during an exceptionally low tide to the enormous mussel beds of Ellison's Scaur off Dubmill Point. Black shells stretched out in all directions over the lumpy terrain, in sheets and metre-high domes, dotted here and there with the orange bodies and curled limbs of predatory starfish, *Asterias,* their suckers slowly, inexorably, pulling the mussels' shells apart. These mussels are part of the Solway's undersea for most of the year; the Ross worms, *Sabellaria spinosa,* which build tubes like the intertidal honeycomb worm, *Sabellaria alveolata,* live even deeper. Buried in the sand at or below the low-tide mark are animals like the razor shells, *Ensis,* and heart urchins, *Echinocardium cordatum* (also known, less romantically, as sea potatoes). But what cannot be seen, cannot be understood or cherished. And what is abundant one year may decrease the next: the Ellison's Scaur mussel beds have temporarily all but disappeared.

Ships longer than fifty metres going up the Solway Firth to Silloth must take on a pilot from just off the port of Workington. It's then a ninety-minute voyage, with known and occasionally unknown hazards, and the pilot and ship's master must negotiate the English Channel – limited to the north by the Workington Bank – then pass through the Maryport Roads, look out for the Solway Buoy, head north to north-west by Corner Buoy just off Ellison's Scaur, followed by a wiggle north-east towards the Beckfoot Buoy and into the Silloth Channel. At low tide, the entrances to English ports like Workington, Whitehaven and Silloth are inaccessible – the water inside the docks is retained by dock gates, which open at high tide. The entrance to Silloth's outer Marshall Dock is at right-angles to the tidal flow: difficult at any tide but especially so when the Solway's biggest spring tides – potentially nine to ten metres at high water – are running. Entry must be at high water, when the tidal flow is at its minimum, but 'there's no slack water at Silloth', Ed says. He has to make a passage plan for each ship's master, and the pencilled diagrams that he shows me are graphic explanations of the ship's potential movements. 'We stop the ship three cables off the entrance and wait for the tide to drop off, then I'm aiming at West Beach, crabbing in – the counter-current off the entrance is pushing the stern in the opposite direction even as the bow is inside the dock.'

One time I watched a visiting pilot bring in a ship. Its port bow was pushed against the wall of the entrance and there was a puff of red sandstone dust; then it was pushed back towards the starboard side, just missing that wall by centimetres. The pilot told me later, 'If there wisnae the counter-current it would be easier – the bow was pushed to the south and I thought "bugger!". You can bring the same ship in half-a-dozen times and it doesnae react the same way – it all depends on the tide, the wind, or if it's in ballast . . .'

Entering the dock presents different problems, too, Ed told me. 'When you enter a narrow channel, there's a hydraulic effect, the water speeds up under the ship. It's asymmetrical and unbalances the ship. Then you enter New Dock, which is the shallowest point, and the hydraulic effect increases. The flow underneath causes the ship to "squat" – it sits lower in the water because the pressure has dropped.' New Dock, the inner dock at Silloth, isn't huge, and there may be other vessels there already, ranging from scallop boats from the Scottish side to cargo vessels, taking up space along the quays. 'We get into the dock and then we have to turn the ship. The captain needs to tell us how the ship behaves when going astern – it's an effect of the propellors – whether the ship tends to swing port or starboard.' Tricky enough on a calm day, but even harder when there's a strong wind and the space is limited. Tricky at night, when the two onshore leading lights must be lined up for the correct angle to approach the gates. And even trickier when the vessel being piloted is the *Zapadnyy*.

The *Zapadnyy* carries molasses, for use in animal feed. Transporting molasses is difficult. It has to be kept warm by heating coils so that it remains fairly fluid. But the Production Manager from Carrs' factory on the quay told me the molasses can also undergo an exothermic Maillard reaction – like caramelising – when 'it turns to coke. I've seen people having to take a jack-hammer to it to release it'. *Zapadnyy* is my favourite ship on the Solway, a ship whose erratic behaviour collects stories. I first heard about her in connection with some emergency welding that had to be carried out on Silloth dock, and all the Silloth pilots have stories to tell about her: 'She's unmanoeuvrable!'; 'It's anybody's guess which way she's going to go'; 'She's got such a broad beam. It's like trying to steer a coracle!'

One afternoon in January *Zapadnyy* is due in to Silloth, and Workington's harbour master, Russell Oldfield, has offered me a trip out to the ship on the tug and pilot boat *Derwent*. At breakfast time, I watch *Zapadnyy*'s red icon on the live shipping website as she makes her way up the Irish Sea, turns east into the Solway, and slows and anchors off Workington just after 9 a.m. The Marine Traffic website gives all her details, her tonnage, her deadweight (weight including cargo), her length and breadth, and more: a ship-spotter's dream. She's an old ship, built in 1988, and she's registered in Belize – but I know that she has a crew of Russians and Ukrainians who, despite political differences, apparently coexist amicably in this confined space. She needs to wait off Workington for the late afternoon high tide, but the wind has got up and the rolling brown waves on the Firth are streaked with white. The forecast is bad – a north-westerly wind, force 5 or 6, increasing gale force 7 later in the afternoon. Tim Riley, pilot and Silloth's current harbour master, phones me at midday to say that none of us will be going anywhere today. I watch online as *Zapadnyy*'s icon swings round to the south-west and increases speed, direction 200°. She will sit out the gale elsewhere, heading out into the Irish Sea and then sheltering off the Isle of Man overnight: no shore time for her crew tonight.

The next day the wind has moderated and I see that the tanker is back off Workington, speed 0.0 knots – her crew waiting, again. At the port I'm directed towards the far end of the dock where the *Derwent* is moored. The deck lies fifteen feet below me, for the dock gates are open and high tide is still two hours away. Russell Oldfield is washing the deck. He directs me to the far side of the dock next to the RNLI station, where there are fretted metal steps. Oily water sloshes to and fro beneath them, and green and grey shadows shift and shape-change in this under-world amongst the concrete piers. Out past the breakwater the swell is noticeable, and the view from the windows is bleared and distorted by blasts of spray as the bow plunges and slams into the waves. *Zapadnyy* is visible now, idling under motor, her bow pointing west, but she is of course expecting us, and as her captain and our skipper, Phil, talk on the radio, smoke puffs briefly from her funnel and she slowly turns around, wallowing gently, waiting. She has come up from Avonmouth, where she took on her cargo of molasses from a bigger tanker, but she is not fully laden for her red-brown hull is partly visible. Phil slowly brings the *Derwent* along the starboard side, in the lee of the north-westerly wind; two of the crew are waiting to welcome Tim, and he very quickly climbs the short rope ladder and is aboard. *Zapadnyy*'s crew, rolling up the ladder, smile broadly and wave. In seconds, we're going astern and the tanker is under way: the meeting of ships and humans of different nationalities, briefly united by the sea, has been terminated.

We swing away, and now Russell sits at a computer and Phil, watching another screen, steers us along a preordained course towards the west. All the ports are constantly silting up with sediment that has been redistributed by the Firth, and where the river water meets the sea, as at Workington and Maryport, it drops its load – whether sediment or tree trunks or plastic chairs and garbage. Paying for a dredger to come and remove the débris – and dump it in designated 'Soil Grounds' in the Firth – is expensive, and not something that harbour masters want to do too often. The effect on the seabed is probably not good, either. Under the terms of its licence to dredge, the Port of Workington has to check the movements of that dumped material, so now *Derwent*'s sonar is measuring and recording depths along the designated Grounds. I stand out on the deck to watch, and feel, the rise and fall of the waves. A guillemot takes off in a flurry of wings and running feet, and the sun's rays suddenly fan out from a hole in the cloud, gilding the surface of the sea.

Not long afterwards I drive north-east along the coast road up to Silloth, catching glimpses of *Zapadnyy* in the distance as Tim guides her through the channels. A smirr of rain hides her as she passes in front of Criffel, but soon she is slowly passing the harbour entrance, losing way, idling – waiting. Someone comes down from the harbour office to check the tide gauge by the dock gate: he waves at me and shakes his head – not enough water yet for the tanker to make her entrance. Fifteen minutes pass, and then her masthead riding lights appear above the wall and she approaches gently, gently, delicately turning in to the difficult entrance, sweetly gliding through the outer dock and setting the anchored shrimp boats swaying; gently through the narrow entrance of New Dock, and safely into port. It's almost dark now, lights from the warehouses glittering on the water and scattering as *Zapadnyy* manoeuvres to her own special place. Her entrance was perfect.

But it has not always been so trouble-free. One of the photos on the Marine Traffic website shows her elsewhere with a spectacularly damaged bow. And here at Silloth, where wind and counter-currents at the harbour mouth make for a very difficult entry, she has on occasion rammed the dock wall – the incident when emergency welding was required – and, worse, on a visiting pilot's watch, grounded outside the harbour on a newly accumulating sandbank. On that occasion, she had to stay put (much to the delight and interest of ship-spotters close by on the shore) until she could be hauled off on the night's high tide.

This evening, though, she has arrived quietly and calmly. The harbour staff in hi-vis are waiting on the quay, ready to secure the ropes and hawsers. She is declared 'all fast'. The crew wave to me again, and then busy themselves with their well-practised tasks. They will spend tomorrow at Silloth – there should be time to go ashore – and then Tim Riley will guide them back down to Workington on the falling evening tide. Meanwhile, the pipe will be connected and warm molasses will start to flow, to be stored then mixed in Carr's harbour-side factory to make Crystalyx for sheep and cows. *Zapadnyy* may be difficult to steer but, as pilot Ed Deeley says, 'She's got a very competent bridge team as well as a very good echo-sounder! And she's the only vessel that manages to keep molasses in a good state.'

*

In November 1796 Captain Joseph Huddart of Allonby, a competent seafarer and surveyor himself, gave a very technical paper to the Royal Society concerning refraction of distant objects by the atmosphere – essentially, about mirages.[10] From Allonby on the Cumbrian shore he had been able to see Abbey Head, about twenty miles distant and across the Firth in Galloway, apparently floating above the sea. He could also see the 'dry sand' of the Robin Rigg sandbank about four miles away; at that state of the tide the visible sand was three or four feet high. The top of the sandbank was known until fairly recent times as a place where shelduck would congregate. In 2010 E.On's sixty-turbine wind farm built on the Robin Rigg and other nearby sandbanks came online and gradually the above-sea section of the Robin Rigg sandbank disappeared.

This is unsurprising given the Solway's tides and the nature of its seabed. Before and during the construction of the offshore wind farm at Robin Rigg, several important surveys looked at the geology of the seabed and shore, and the behaviour of the tides. Much of this interesting information about the soft sediments of the seabed stresses the mobility of the superficial sediments of fine to medium sand and sandy mud, material that was scoured out by glaciers and deposited by rivers and tides.[11] In the region of the wind farm the seabed is in fact a series of banks (Dumroof Bank, Robin Rigg, Two Feet Bank and Three Fathoms Bank), orientated by the currents in a north-east–south-west direction and made of fine-grained sediments '. . . ranging from laminated sands, silts and clays to organic silts and clays'. So we begin to get a picture of the bed of the Firth and, given the heights and speeds of the tides that funnel in from the Irish Sea it's easy to see how the sediments can get moved around. At neap tides, the tidal range (the difference between high and low water) may be as little as three metres, but for the largest spring tides, the range can approach ten metres; the range for mean spring tides is about eight metres. That's a lot of water that has to shift in and out of the Firth every twelve hours or so – and the speed for mean spring tides has been 'shown to be around 1.9m/s between Dubmill Point and Southerness Point and around 2.4m/s at Annan to Bowness' according to a survey carried out in 2010.

The day that I waded across from England to Scotland on the Bowness 'wath' or crossing-point at low tide, it was unnerving to feel the bed of the Firth shifting beneath my boots, and whenever I paused to regain my balance a hollow quickly formed downstream of my feet. Scouring – the sucking away of sediment to form a hole – was a well-known phenomenon even back in the 1860s, when the railway viaduct was built just west of Bowness. Just as with the wind farm, the seabed was surveyed beforehand and 'In the trials and borings that were made it was found that the bed of the Solway is composed of very strong coarse gravel, interspersed with boulders, and on the top of this gravel there is generally from five to six feet of sand, *which is constantly being shifted by the currents of the Firth*' [my emphasis] (*Whitehaven News*, 11 February, 1869).

Later, after the viaduct was so severely damaged by ice floes in 1881 (Chapter 1), the Board of Trade Inspector, Major Marindin, recommended that 'dolphins or fenders of some kind must be provided to prevent the ice from touching the piers . . . such things are apt to cause a scour in the bed of a stream, and their foundations in future must be carefully watched'. More than 130 years later, in October 2016, a side-loading hopper barge spent some time at Robin Rigg wind farm dumping rocks 'for scour protection' at the bases of five of the turbines.

By 2015 part of the Robin Rigg sandbank had become very unstable, as the adjoining channel had deepened. Shrimp trawlerman Danny Baxter (Chapter 8) told me how he had noticed the changes in the channel: 'Going over to Wigtown, years ago, it was about thirty feet deep, but it's all washed away now, down to seventy feet deep.' And so it was that two of the wind-turbines had to be taken down. A jack-up vessel, MPI *Adventure*, steamed onto the scene in October 2015 and whenever the weather allowed, she put down her legs and her cranes went into action. Whenever the wind got up, *Adventure* paused or headed back across to Belfast until the working window of the next neap tide. I watched from high up at the Senhouse Roman Museum at Maryport on one occasion when she returned, and she was an imposing sight, travelling fast for such a giant.

But out of those hidden changes on the seabed came another previously hidden story. Photographs that E.On shared with me showed that the subtidal section of the monopile, the cylindrical base of the column, had become an artificial reef that was so densely, darkly, encrusted with marine organisms that there was no free surface. E.On had also previously commissioned a series of surveys of benthic (bottom-dwelling) fauna, sea mammals and sea birds both during and after the construction of Robin Rigg. Using drop-down video cameras, Dr Jane Lancaster and colleagues at Natural Power found an abundance of marine life colonising the bases of the turbines and substations after only four years.[12] The intertidal sections were coated with typical rocky-shore organisms such as barnacles, limpets and green algae with a 'dense carpet of edible mussels around the low water mark'. The mussels continued down below the low-water mark, 'providing a rich feeding ground for starfish, as well as green sea urchins which feed on the anemone-like colonial animals [hydroids] that live on the mussel shells'. On the bases were 'forests of brightly coloured plumose anemones' as well as hydroids, and shore crabs and hermit crabs crawled around the foundations and on the seabed. Fish like whiting were seen swimming around the piles. Above the surface, the cormorants that line the rails of the platforms like crotchets on a stave are probably a sign that fish – in the absence of any trawling round the wind farm – are abundant.

At a meeting of 'interested parties' to discuss the future of renewable power generation in the Solway Firth, there was enthusiasm about these artificial reefs and the new species of wildlife they encourage into the Firth. But, to laughter around the table, a member of a conservation organisation muttered glumly that putting turbines in the Firth was like 'planting trees in a flower meadow'.

*

Harbour walls and jetties, offshore wind farms, the embankments of the Solway viaduct, potential tidal power lagoons – all these are influenced by, and in turn have an influence on, the way the Solway is 'made' and remade. As I write this, the Greenland ice cap is melting at a formerly unimaginable speed and portions of the Antarctic glaciers are moving faster towards the sea, lubricated by meltwater. It is certain that the global sea level will rise – and the Solway Firth will widen once more, and reconfigure itself, with consequences for every living thing along and in it. We shall be able to watch and record this happening, of course. With the help of satellites and NASA's Earth Observatory satellites we'll be able to pore over a time-sequence of the changes along the Firth. If only there had been this technology when the glaciers retreated! We'd have been able to shift the on-screen slider left or right, comparing the low-tide images through the decades and centuries and millennia – watching the shape-shifting interface of sea and land, and the growth and decline of trees and bogs. Laser Light Detection and Radar (LiDAR) would have measured contours with ten-centimetre accuracy. Satellite instruments would have measured the changing temperature of the waters, and we could have noted where the phytoplankton bloomed, and made good guesses about the spawning of shoals of herring, the breeding of thornback rays and the migrations of andromadous fish like sparling and salmon; we would have been able to spy on the migrating flocks of geese and waders too. They are all vertebrate species with individual and collective memories of place, and influenced by the seasons and the tides.

But the invertebrate animals and algae, the important 'eco-system engineers' and a source of food, would have been key to the influx and survival of those vertebrates. When did the sea anemones, lugworms, barnacles and heart urchins arrive and colonise the rocks and sand? They could have come at any time after the ice melted, as long as the salinity and temperature suited them, because they are all dispersed by planktonic larvae. Not so *Corophium*, however, whose young develop in their mother's pouch and only leave home to dig their own burrows when they can crawl and swim. During the last Ice Age the mudshrimps must have hung on in estuarine mud and other unfrozen refugia further to the south, then gradually, very gradually, travelled north on the currents around the Irish Sea, plucked from their colonies by the tides – and surviving to breed only where the muddy conditions were 'just right'. There would have been space for their colonies to expand as the coastal margins grew. Their numbers and distribution would have waxed and waned with the availability of mudflats. Mudshrimps would have been exploring the Firth's edges at the same time as humans.

Comparing the genetics of different populations from around the coasts would give us a clue to the Solway mudshrimps' origins, but of course the research has not been done. Unlike those of Canada's Bay of Fundy, ours are not deemed sufficiently important.

Fundy's *Corophium* are a phenomenon, in places as densely packed as 50,000 per square metre and the main food source for the enormous flocks of sandpipers that pass through in the autumn.[13] *Corophium* is widespread around the coasts of the north-west Atlantic yet – unlike its other co-inhabitants of mudflats such as the snail *Hydrobia* – it isn't found in Iceland or Greenland, which would be expected if the mudshrimp migration across the Atlantic had involved those stepping-stones. Tony Einfeldt and Jason Addison are among the most recent Canadian researchers to wonder about those origins and they persuaded colleagues and other scientists on this eastern side of 'the pond' to send them alcohol-preserved specimens of *Corophium* for analysis and comparison of mitochondrial DNA.[14] Without going into detail here, the conclusions of their carefully argued and well-evidenced paper show that the Fundy populations bear so many similarities with the north-east Atlantic populations, and show so little genetic divergence within themselves, that it is likely that the Canadian mudshrimps were essentially 'brought over' to North America. It seems that *Corophium* can survive inside blocks of ice but, because their death rate is high, causing their populations to decrease during winter, the authors conclude quite reasonably that dispersal by 'ice-rafting' is highly unlikely. Instead, they concentrate on shipping: European colonists and traders established settlements in northern Canada and the Bay of Fundy area during the early 1600s and, before ballast tanks were introduced in the 1860s, the buoyancy of ships was controlled with semi-dry ballast, 'i.e. rock and sediment loaded into damp holds that was collected from heaps in the intertidal'. So, because mudshrimps live in the soft sediments of estuaries and river mouths that provide safe harbours for shipping, this 'would likely have enabled the inadvertent movement of many individuals'. What an intriguing idea! In the 1800s, ships from the Solway's estuarine ports of Dumfries, Annan, Maryport and Whitehaven were frequently crossing the Atlantic to Quebec carrying emigrants as well as goods. Could they have been carrying mudshrimps too?

How often ships left the Solway carrying solid ballast is now almost impossible to find out, nor are there records to show where they would either have dumped or collected ballast. Tony Einfeldt, though, sent me a link to historical ballast sites along the east coast of Canada. In the harbour of Richibucto in New Brunswick there are two large ballast islands, clearly visible in photographs and shown on maps as early as 1827. Richibucto became important for the export of timber destined for Europe, and at the same time, 'numerous immigrants, mostly from Ireland and Scotland, arrived on ships that went to Richibucto to load up with lumber. This resulted in the creation of an immigration society to assist the newcomers. In 1832, for example, some 65 passengers came aboard the *Isabella* from the Irish [*sic*] port of Carlisle'.[15]

This probably refers to the (English) Port Carlisle on the Upper Solway, before it was silted up. In a nice coincidence, one of the information panels on the quay at Annan harbour has a painting of the brig *Helen Douglas* in full sail; she was built in Richibucto by John Nicholson, who had worked in his family's shipbuilding company at Annan before he and his carpenter moved to Canada in 1825. The *Helen Douglas* then sailed back from Richibucto to Annan with a cargo of timber for the Nicholson shipyard and she continued to sail the same route for the next fifty years, before – an ignominious end – being sold at Maryport in 1882 for use as a 'coal hulk'. Alan Thomson, Annan's harbour development officer, thinks that river gravel would undoubtedly have been used as ballast, as it was near to hand and plentiful, although he has never seen any written accounts to back this up. The loading area, the lade, frequently silted up and was cleaned out by men with shovels – but it seems the gravel that they shovelled out was used in the building trade, and the loosened mud and silt was flushed out into the river by opening the sluice gates at the north end of the lade. Later, broken red sandstone was often piled on the quay to use as ballast.

We can make guesses about the mudshrimps' arrival in the Solway and North America, but going back considerably further, we don't know when mudshrimps arrived as a species – when they evolved – because the fossil record for amphipods (though not other crustaceans) is very poor. Dr Lauren Hughes, Curator in Charge of Invertebrates at London's Natural History Museum, and an amphipod expert, told me that amphipod fossils only go back to the Eocene period, between 56 and 40 million years ago 'and don't ask why as that is one very big mystery'. Where *Corophium* fits into the evolutionary network of amphipods is also unknown, though it's presumed that benthic – bottom-dwelling – amphipods developed before the swimmers and burrowers.

As for the collective memories of mudshrimps, 1,000 years equates to at least 1,000 mudshrimp generations. Even if they could speak, any putative memories of the past would be a cacophony of Chinese whispers. For several years there was debate and research on whether crustaceans like crabs and lobsters 'felt pain' when they were killed in the kitchen by boiling. Why would they not? These are complex animals with a nervous system and a brain, and an array of sensory organs to help them see and understand where they are in their world. Recent research on the pistol shrimp has shown that its brain contains a structure called a mushroom body, a memory and learning centre previously only found in that other diverse class of the Arthropods, the Insecta.[16] It makes sense for an animal that expends considerable energy in digging and strengthening a burrow, the place where it will live, feed, mate and escape from prey, to remember where 'home' is. We expect this of vertebrate animals, but we know that invertebrates do it too – think of bumble bees, as well as many other insects and molluscs. Crustaceans are no different: Norway lobsters, *Nephrops*, with their large branched burrows, remember how to get home; so do many crab species, and pistol shrimps. A male *Corophium* who enters a female's burrow is swiftly ejected after mating and unwelcome in his neighbours' burrows and, without being too anthropomorphic, it would be comforting if he could find his way back home. That research has not been done, either. But as the contours of the mudflats along the Firth's edges alter, so the mudshrimps and other inhabitants must adapt by moving to the areas that suit them best.

3

Ships and seaweeds

Ships speed past on calm seas, entering or leaving port. A yacht motors, fast, into Maryport Marina; a shrimp boat leaves Silloth, a little erratically; cargo vessels, empty and with red hulls showing, or laden with their Plimsoll lines barely visible, arrive or depart from Workington or Silloth. The movements are speeded up, ships zig-zagging, ships seen from unusual angles. Film-maker Julia Parks captured the shipping movements during 2015 and 2016, using 16mm film and a second-hand Bolex cine-camera.[1] Energetic and enthusiastic, Julia recently returned to live in West Cumbria, and we occasionally meet up along the coast to 'talk of many things, of shoes and ships and sealing-wax', among them. This time we were scrambling over the slippery rocks on Parton beach, enjoying the rows of winkles crammed into fissures, the newly developed platforms of honeycomb-worm tubes, and the swirls of colour in the local sandstone. Julia – like a few people on my low-tide shore-walks – is a good 'finder' and she thinks this ability to notice detail could be the cause, or perhaps the effect, of her work as an artist; in less than five minutes after leaving the car park she had picked up a small slab of rock on which was the perfect dark filigree of a fossil fern. Her final piece of work for her degree course had been filming shipping on the coasts of Devon and south-west Cornwall, partly because she had an aunt in Launceston with whom she could stay, and when she returned to Cumbria she wanted to follow this up using 'animation' (what we used to call 'time-lapse' when we used a similar Bolex camera in the lab for filming the movement of mammalian or insect white blood cells *in vitro* in response to various stimuli). But ship movements on the Solway are not frequent and are dependent on the tides, so 'I'd have to ask if I could leave work so I could go just for a couple of hours to catch a boat,' Julia explained. 'But I met all the amazing ship-spotters – they were really helpful. And animation is a process I really like – you can use it to compress a lot of material.' The result is a three-minute film that portrays the ports and their entrances as well as the comings and goings of boats on the Solway.

The Marine Traffic website[2] shows ship movements in real time, the icons coloured according to type – red for tankers, green for cargo vessels, turquoise for 'tugs and special craft'. There is a preponderance of turquoise craft busying around the wind farms off Walney and at Robin Rigg; cargo vessels are travelling between Liverpool and Irish ports, or heading north; there are passenger ferries crossing from Holyhead and Cairnryan. At this particular moment, despite the wild weather brought by Storm Lorenzo, the lifeboats (turquoise) are thankfully all moored in their ports (circles rather than ship icons with pointed bows). The English Channel around Dover and the North Sea around the Netherlands show packed shoals of coloured icons, but the Irish Sea does not look very busy in comparison. The North Sea around Scotland's east coast is strangely monochrome, dominated by 'tugs and special craft' servicing the wind farms and oil rigs.

In *The Making of the British Landscape*, Francis Pryor has drawn a map which swivels the usual perspective so that we look south from a viewpoint near the Arctic Circle.[3] North is now at the bottom of the page and Shetland is the focal point of his map, a little to the right of Norway – it's a pivot point near the tip of Scotland where the decision must be made: whether to sail 'left' into the North Sea or to head 'right' down the Scottish coast to Ireland. The bulky interiors of Norway, Denmark, North Britain and Ireland are suddenly seen to be unimportant in relation to their edges. It's those edges that mattered in the migration of humans and other living things. Despite the prevailing south-westerly winds and currents, humans could work their way down the firths of the Scottish east coast, or down the sea lochs and islands of the Scottish west coast, and then make a decision whether to stop at Ireland, or shelter in its lee and navigate through the North Channel into the Irish Sea. Pryor's map extends south as far as the coast near Liverpool, and the Isle of Man is seen to be a reassuring stepping-stone. Draw another map with the Isle of Man at the centre, and the Irish Sea becomes an elongated pool bounded by the coastlines of five countries,[4] with its entrance and exit at the top and bottom. If you have some means of self-propulsion – paddles or sails or engine – then depending on the weather and the tides you have the power to decide whether the North Channel or St George's Channel will be your exit. Over to one side, the Solway Firth is a route deep into the interior, or out of it, for escape, adventuring or proselytising.

For too long, I thought of early humans in terms of them migrating north and west, then arriving at a coast, looking out at the water, and saying 'oh dear!' But, amongst our many characteristics as a species, we are inventive and willing to experiment. The boats hollowed out of oak logs that are displayed in Dumfries Museum show that the Bronze and Iron Age locals were prepared to utilise the water to move around. Whoever paddled the log-boat that was found at Lochar Water might have regularly ventured out from that base onto the Firth to fish, or he might have been swept there by the tides. Any form of marine transport would have offered freedom, food and new land; the edges of the Upper Solway were muddy or marshy, but there were sandy and pebbly beaches too.

St Columba and his twelve companions sailed from Ireland and reached Iona in a boat made of cowhide, a coracle, in AD 565. How St Ninian reached the Isle of Whithorn on the Solway's Galloway coast at the end of the fourth century is not at all certain. Indeed, very little is certain about Ninian, and it's possible that the Venerable Bede's account of his life, written several hundred years later, was actually about someone else entirely. But Whithorn, with its Casa Candida, the 'white house', became one of the early Christian centres, and has been a place of pilgrimage – by sea and land – ever since. As we have seen, the

Romans and the Vikings used the Irish Sea as a thoroughfare and a gateway into the lands each side of the Solway Firth, and in 1300 Edward I called up a naval flotilla from the south coast of England to help him in the War of Independence against his northern enemies.

The Firth became busier with shipping, so that on any day you might have looked out and seen a range of skiffs and fishing boats and two- or three-masted sailing vessels. Boats of all sizes were being built or mended in coastal ports and up the rivers. At Maryport a gridiron is still visible near the mouth of the River Ellen and next to the Maritime Museum; its parallel rows of wooden timbers supported on iron struts are now only a few feet above the muddy river bed, but at one time boats would have been floated onto the grid and men would have worked beneath the hulls. In an oil painting from the mid-1800s by well-known maritime artist William Mitchell, the herring fleet is seen leaving Maryport: seventeen or eighteen boats, their brown sails bellying in the wind or with their free edges luffing as they are hauled to the top of the mast. The wind is from the north-west and the boats are tacking across each other's bows, past the harbour wall and lighthouse. The delight of Mitchell's paintings is in the movement and the details, the gulls squabbling over a piece of flotsam, the blue jersey of the rower in a skiff.

Boats of all sizes were entering and leaving the Firth; the quickest way to travel between England and Scotland would have been across the Firth, and communication and exchange by that route would have been the norm. The Upper Solway must have been crowded when the tide came in, especially around Annan – James Irving Hawking's book, *The Heritage of the Solway Firth*, is a precious store of photographs and paintings of ships and shipbuilding up until the 1970s.[5] The steamship *Jennie,* white smoke trailing from her funnel, tows three elegant topsail schooners down the River Annan in 1905. There are fifty-one trawlers, thirty whammel boats (for seine-netting salmon), thirteen herring boats and twenty-four 'small boats' at Annan in about 1900; photographs show the long row of the rebuilt piers of the Solway viaduct in the background. Goods like coal, haematite and wool are imported or exported along the Firth. Schooners and barques are setting off to ports around the Irish Sea or further afield. The young Joseph Huddart and the Allonby herring boats head into the Irish Sea and out of the North Passage to the Outer Hebrides. Scandinavians, English and Scots travel by sailing boat and barge along the canal between Carlisle and Port Carlisle to embark on steamers that will take them to Annan and down the Irish Sea to Liverpool, where they will board ships to carry them to a new life in North America and Canada. Shipowners in Whitehaven grow rich on the triangular trade in enslaved people and rum and sugar.

In the twentieth century, between 1950 and the mid-1970s, Solway shipping photographs are of coasters with cargoes of timber, cement, wood and fertiliser; granite blocks quarried near Creetown are loaded at a quay. In 1943, Garlieston in Wigtown Bay is the secret site for test-ing different designs of floating dock for the 'Mulberry Harbour' landings on D-Day.

*

In a Firth with 'changeable depths', sandbanks and occasionally fast tides with a great tidal range, there will always be shipwrecks, and for me the stories of two, in particular, have been partly brought to life.

A few years ago I finally found what I had searched for on several occasions, the vestiges of a ship's keel – on Ship's-keel Scaur in Allonby Bay. Its timber was hard as iron, and the keel (if that is what it was – its profile had been much transformed: holed, distorted and overgrown) was home to a variety of marine species, a microcosm of the animals and algae on the shore. Honeycomb worms, *Sabellaria*, had congregated to build a low-rise block of their sandy tubes; the pale grey cones of barna-cles dotted the dark wood and the rusty iron chain, their kite-shaped trapdoors closed until the tide returned. There were several tightly clinging limpets, and moving, grazing winkles, *Littorina littorea*. Predatory dog-whelks, *Nucella*, their shells dirty-white or striped with brown, had laid mats of their orange vase-shaped eggs underneath the overhanging timber. There were deep red and dark green beadlet anemones, *Actinia*, in intertidal mode as squat globes with their tentacles tucked inside their stomachs; and thin flat 'leaves' of the sea lettuce, *Ulva*. Footprints show that wading birds had found shelter next to the timber. Although six or seven metres long, the keel blended into its surroundings on the scaur, providing yet another stable surface against the shifting sands; its encrusted chain bled oxidised iron. It was also scarred by deep rectangular excisions, caused by present-day hacksaws – someone who joined one of my shore-walks told me that he and his friend had been extracting copper nails, and he showed me a wedge-shaped nail, several centimetres long and tapering at one end. The nail, the keel, were old, that much was obvious – but how old? Which vessel did it support, and how did it come to be cast up in Allonby Bay?

Allonby's most famous shipwreck was of the barque *Hougoumont*. Bound for Liverpool, she was driven north to Maryport in a storm early in 1903; she dragged her anchors and ran ashore at Allonby and her cargo, which included '32,000 cases of tinned pears and peaches, and 24,000 cases of salmon', was much appreciated by local people. Her story is well told and illustrated in Peter Ostle's blog and book about Allonby's history.[6] But she is not the source of the ancient keel, because her battered hull was eventually towed to Maryport to be repaired. At Maryport Maritime Museum is a black wooden board bearing the hand-painted 'List of Lives Saved by the Maryport Life-boat between 1888 and 1944': '1903 Feb. 27th Ship "Hougoumont" of Glasgow landed. Lives saved 25'.

Wanting to find out more about the ship's keel, I put a small notice in the local news bulletin, the *Solway Buzz,* asking for information, and several weeks later John Whitwell, a volunteer at the Maryport Maritime Museum, phoned me, wondering if he might have found an answer. He had been hunting through the archives and found a notice and a marked-up Admiralty Chart that suggested the intriguing possibility that the keel might be from the 1,180-ton barque *William Leavitt*, which was stranded on Barnhourie Bank then refloated, but finally wrecked just off Dubmill Point in November 1888. There is a long article about the *William Leavitt*'s plight in the *Maryport Advertiser* of Friday, 30 November 1888.

EXCITING SCENES ON SUNDAY

On Friday afternoon the Norwegian barque, William Leavitt, of Laurvig, 1181 tons register, Captain Gude, bound from Quebec to Greenock with a cargo of timber, was driven into the Solway Firth through stress of weather and anchored in the roads off Maryport and Workington. The captain afterwards went ashore at Workington for the purpose of telegraphing to Greenock for a steam tug, and as a very strong wind was blowing and a high sea running he was unable to return to his vessel. When night came on the gale, which had been blowing all day, increased in violence, and the barque began to drag her anchors. About 11 o'clock the steamer *Plantaganet*, from Liverpool for Maryport with a general cargo, came off St Helens, and seeing the 'flare up' light, the captain altered his course and bore down upon the craft. On getting near he noticed that the vessel was drifting inshore, while the crew were running about with lighted torches, being apparently panic-stricken. The steamer sailed round the barque and came quite near her on the lee side, but the captain could get no intelligent response to his frequent inquiries. This circumstance is now explained by the fact that the barque's crew were all foreigners and unable to speak English. The *Plantaganet*, however, stood by until the ship drifted past Maryport and came off the Salt Pans, situate about 2½ miles higher up the Solway. Although it was a fearful night and the steamer had enough to do to live in the surf, the captain is of an opinion that if he had got a rope from the barque before she passed Maryport, he could have held on until assistance arrived. The first indication that a vessel was in need of assistance appears to have been observed at Maryport at about twenty minutes past eleven. At that time the pilot boat *Ally Sloper*, owned by Mr. W. Walker, junr., and manned by John Robinson (master), James McAvoy, John Messenger, and John Byers, was on the lookout for a steamer expected by that tide, and seeing the light the men got into the boat and went out. After getting outside the harbour the frail craft was in danger of being engulphed at any moment by the angry waves, but the men held on until they came within speaking distance of the barque in distress. Just then the pilot boat's foresail and mizzen were carried away and the crew placed in greater jeopardy than those they were attempting to rescue. When in this plight they fell in with the *Plantaganet* and all having after with great difficulty got safely on board with the boat secured behind, Robinson, the pilot, took charge of the steamer, and made for the harbour. When running up against the wind the steamer passed both the steam-tug *Senhouse* and the lifeboat going after the barque. About this time the Norwegians abandoned the vessel, and she ran ashore at Dub Mill, while they continued to drift about in an open boat until they were picked up by the tug . . . The men, 16 in number, were landed at Maryport and taken to the Coffee Tavern at about three o'clock on Saturday morning. They were all much exhausted and two were suffering from slight injuries received while getting into the boat to leave the ship. The lifeboat did not arrive in time to render any assistance, but she remained close to the wreck for some time after the tug returned to make sure that none had been left either on board or near the barque . . . The coxswain states that the lifeboat behaved remarkably well in the surf at Dub Mill. The barque is most likely to become a total wreck, but it is probable that most, if not all, of her cargo of 1,800 tons of timber, will be got ashore. On Saturday forenoon the master attended at the Custom House to make his deposition. The shipwrecked crew were taken charge of by Mr. J.B.Mason, Vice-Consul for Sweden and Norway.

On Monday evening the crew boarded the vessel and brought their clothes ashore, but as yet no attempt has been made to float the ship.

A good deal of the wreck was washed upon the beach between Silloth and Allonby.

The article brings so many different aspects to life – the human stories, the drama and the range of boats involved, the end of a transatlantic voyage, the cargo of yet another load of Canadian timber to accommodate the enormous amount of shipbuilding that was going on in Britain. Around the Solway there were shipbuilders on all the major rivers, and at Whitehaven and Maryport (where Joseph Huddart had his brig *Patience* built[7]). Joseph Ismay had a timber business and a shipbuilding yard at Maryport in the mid-1800s and it was his son Thomas, born in Maryport, who bought the Liverpool-based White Star Company in 1868. One of the company's ships was the *Titanic*.

A few years prior to the *William Leavitt*'s wrecking, William Mitchell had painted 'FLORENCE and GIMELLO', 'The steam tug *Florence* leading the 598 ton Italian barque *Gimello* into Maryport, 18th October 1883'. The dramatic oil painting is in the Maryport Maritime Museum, and it shows the three-masted barque, sails furled, heeling to starboard and the tug, black smoke streaming from its funnel, heeling to port; both are scarcely visible among the short green seas which are lit by the light of the storm. The vessels are close to the wooden pier, but they are not yet there or in its lee. The painting strongly conveys the sense of dread and danger, and gives us an idea of the conditions on the Firth during which the bigger barque, the *William Leavitt*, ran aground.

There is also a self-portrait of William Mitchell (1823–1900) in the museum, painted in 1899. Most of Mitchell's other portraits are rather lacking in life, but he clearly knew himself better than his other subjects: the lower part of his broad face with its long, strong nose is hidden by a bush of white beard and moustache, his forehead by a soft-brimmed black hat, but he has a powerful and rather accusing stare. He arrived in Maryport, from County Down, when he was seventeen, one of many Irish migrants who were fleeing the famine at that time, and found work in the Engine Works of the Maryport & Carlisle Railway, painting the company's coat-of-arms – its logo – onto the carriages and locomotives. He also started to paint maritime scenes, but he didn't leave his job (by then he was Foreman Painter) with the M&CR for another twenty years, by which time his paintings had become popular and he was receiving many commissions. His obituary in the *West Cumberland News* notes that 'He was very industrious, and according to the records which he kept, had painted over 10,000 pictures.' He was also industrious in other ways: he had had three wives and at least twenty children, and was contemplating marrying again, by the time he died.

What happened to the 'good deal' of the *William Leavitt*'s wreck that was washed up on the stretch of shore between Allonby and Silloth? Perhaps even now there are family heirlooms, of cutlery or brass, sitting on mantelpieces or tucked away in drawers, whose stories have been passed down from great-grandparents. And could a wooden keel from the end of the nineteenth century still survive onshore? There is a model in the Maritime Museum of the four-masted *Peter Iredale*, launched from Ritson's yard in Maryport in 1890, and wrecked on the Oregon, USA, coast in 1906. According to the caption, 'she became a tourist attraction for over 80 years, with her timbers still visible today'.

In the early 1900s Allonby was also, surprisingly, a site for ship-breaking, a business operated at the southern end of the bay by the Twentyman family. Photographs from that time show the mastless hulks of ships being towed to the beach. John Whitwell showed me a black-and-white photograph of the steam tug *Florence,* in a less heroic rôle, towing a de-masted hull towards the beach; people are watching from the shore – several women in long dresses and hats, one with a small child on her hip; three men and a boy sitting on an upturned skiff; a group of men and boys up on the deck of a beached hulk. There are other photos, of the part-broken hulls towering over the shore, and of children standing on a plank at the water's edge by the wooden hull of the *Prince Victor.* Ships on the shore, stranded and broken ships. The scenes in the black-and-white photos seem unreal, but it was nevertheless a very practical business. Ships were broken up over a period of many months, but their timbers and metals were salvaged and reclaimed. There are many houses in Allonby that have iron-hard oak lintels above the doors and windows. And my friend Ronnie Porter phoned me recently, having remembered something else from his boyhood: 'We used to say "We'll see you at the logs!"'' They were not logs, he said, but a pile of ships' keels, stacked at the back of one of the fish yards.

Whatever the origin of the ship's keel at Ship's-keel Scaur, its story will never be truly known and soon the keel itself will be no more than a memory. In early 2015 it was about seven metres long, trapped firmly on the Scaur between granite boulders and by a part-buried, heavy anchor chain. The chain was a marvel, its rusty surface crusted with barnacles, *Sabellaria* tubes and green *Ulva*. At one end it held the keel, but the other disappeared down into the sand and pebbles: was it attached to something down there, was its rôle to keep the keel in place? I was especially fond of that chain and its unknown story. Three years later, though, part of the keel had vanished and the chain had been broken off and was now barely visible in the sand; presumably to help find it, someone had jammed a red traffic cone beneath the remaining timbers. A few months later, the keel itself was partly submerged by drifting sand and the boulders had been covered, and in the summer of 2019 I only found it by chance. The Scaur was now almost hidden by smooth sand, and the keel was in two well-separated pieces, battered and fragmented, barely visible. I feel as though I've lost an old friend. Bizarrely, the traffic cone was exposed again, lying in a scour-pool.

I became acquainted with a different wreck in October 2019, when the SCAPE archaeologists were again working with Nic Coombey and the Solway Firth Partnership[8] to survey wrecks along the River Dee. This particular wreck, of the *Fauna*, had probably been beached on purpose, abandoned because its wooden timbers were no longer useful. John and I walked from Kirkcudbright along the road on the Dee's west bank to reach Gibbhill Point, where the hulk was lying tilted on the shingle below the narrow strip of merse and grass. Many of her planks were missing or broken, but those that remained were rough-grained and spotted red with rust. She had been a sturdy vessel, about twelve metres long – but perhaps not so sturdy because, counter-intuitively, many of the planks above her keel were perforated with rows of holes, each three or four centimetres in diameter. Steve Liscoe, the marine archaeologist on site, explained that *Fauna* had been a 'wet-well' boat: her hold had been like an aquarium, where seawater entered to keep her cargo alive. The cargo would originally have been shellfish, but whether oysters or scallops or even mussels was not yet known. I like to think her name reflected her importance in storing marine animals, invertebrates; it's highly likely some mudshrimps could have inadvertently been stored there too. Steve thought that she was originally a Belgian boat; it seems she had then worked out of Fleetwood on the Lancashire coast, after which she had been brought to Kirkcudbright to trawl for flatfish in the Solway. But a larger boat had crushed her when she was tied up in the port, and so she was abandoned at the edge of the Dee, dumped at the top of the tide so that her engine could be removed. The poor *Fauna* is a palimpsest of her likely uses – she seemed to have had a mast and sails at some stage, then been modified for an engine; her stern still retains the metal frame for a trawl. The holes in the hold had been partly blocked by slabs of concrete ballast, probably so that an engine could be installed. Mud, fraying timbers, indeterminate rusty metalwork, tattered deck-planks – she was not going to last very much longer, so SCAPE and their volunteers were recording and photographing as much as they could. Her anchor chain was like a dark, flaking vine coiled in the hold, rust spalling off in sheets.

*

The Solway's commercial ports are now only at Workington and Silloth. Kirkcudbright has Scotland's second-largest fleet of scallop-dredgers; and today only small numbers of trawlers, and lobster- or crab-boats work out of harbours like Whitehaven, Maryport and Harrington. These days, as Julia Parks found when she was filming shipping movements, there are long periods of inactivity.

When I met the manager of the port of Workington, Jeremy Lihou, in 2015, he talked about the changes that had occurred. 'The local community had got used to looking at the port as a failing entity,' he told me. 'They say things like, "You used to be able to walk across the dock on ships".' There were approximately 150 people working here at its peak, as a core workforce. During British Steel's time in Workington, the port shifted 1 million tonnes of iron ore annually, but after British Steel went into decline, the port suffered badly and eventually the local authority, Cumbria County Council, bought it for £1 in 1975. By 2002, Jeremy said, all traditional traffic had disappeared – it had become a dedicated coal terminal. In 2005 a concerted effort by the various local councils and other organisations, including the Nuclear Decommissioning Authority at Sellafield, pushed ahead and provided funding to make the port a container base – with a massive yellow Liebherr crane for picking up containers, and a large area of concrete hard-standing for container storage. Workington was marketed as a 'multi-modal hub' – sea, rail and road, all interconnecting. But the optimism about container traffic, and work for the proposed new nuclear power station at Moorhouse, Sellafield, has evaporated. Now the commodities that are handled by the port are sawn timber, chemicals like iron silicate, cement and fuel oil. Tree trunks and bark are off-loaded for the nearby Iggesund paper-board factory and piled in mini-mountains around the quays. On occasion, piles of plastic water-pipes for United Utilities' new water-pipeline in West Cumbria have arrived, and tanks of salmon smolt from the hatchery at Ullswater have been exported. More spectacular were the large diesel locomotives being lifted ashore from a Spanish vessel. Twin-hulled service boats for the Robin Rigg offshore wind farm, trawlers, the tug *Derwent*, dredgers and tankers: the ships that come and go currently amount to 150–180 per year.

At Silloth, 50–60 coastal vessels per year bring in logs, salt, lime and, because Carr's fertiliser and animal feed businesses are here, grain and fertiliser chemicals, and the *Zapadnyy* brings in molasses. Throughout the scallop-dredging season, the Troon-registered boats come in to discharge their catch. Silloth and Workington – unlike big ports such as Liverpool – can only be entered during a couple of hours each side of high water: thereafter water is trapped in the dock by gates. But at Maryport, home to a small fleet of trawlers, the harbour (although not the yacht marina) is tidal, and across the Firth at Kirkcudbright, where Scotland's second-largest fleet of scallop boats is based, there are no dock gates, so at low tide the boats must wait, moored against the quay, ready to head out to the Isle of Man and the Irish Sea. The rusty red chain-links of their scallop-dredges hang deceptively decoratively along their sides.

Yet despite the mud and the tidal nature of Kirkcudbright's harbour, tankers came up the River Dee between 1956 and 1982 to discharge fuel oil and petrol at Shell-BP's terminal just outside the town. Looking out towards the mouth of the Dee from Gibbhill Point the estuary looks nightmarish to pilot, as the river wriggles east and west, its shipping channels marked by red and green buoys. Near the harbour a final bent elbow must be negotiated, and then an incoming vessel has to be swung right round so as to berth facing downstream with its port side against the dock. The late George Davidson, who died in 2009, wrote an entertaining account of how the harbour was made suitable for these tankers, and the problems encountered by the pilot and the ships' masters: 'Masters [were required] to have their anchors started clear of the hawse pipes ready to let go immediately if required, to assist in swinging round the more awkward marks.' The ship was swung round an anchor in the harbour and then warped in using hawsers from the quay. But 'On one memorable occasion, in darkness and thick fog, as we came in past Gibbhill, I took a 1000 tonner too far to Starboard, and tidied up the remains of the old yair net. On account of the lack of visibility, the pilot boat was close on the quarter and Eddie Parker told me there was so much shattered timber washing around that he thought we were in Bishopton Wood. However, we hadn't touched the ground and the ship was berthed safely.'[9]

The oil terminal was closed in 1982, and the harbour was left to the scallop boats and trawlers. Then in 2018, the cargo vessel *CEG Universe* came in, charging up-river and pivoting neatly next to the quay[10] to take on an unusual cargo – a huge pile of crushed shells from the local shellfish-processing firm, West Coast Sea Products, to be used as the 'grit' in poultry feed in the Netherlands.

There are of course yachts and other leisure craft such as small boats for sea-fishing to be seen on the Firth. At Kirkcudbright they are moored at pontoons and must wait out low water; boats in the marinas of Maryport and Whitehaven are more fortunate, being kept afloat by closed dock gates. Whenever a boat is in trouble in the Firth and the Irish Sea, rescue knows no borders. To the disgruntlement of many, 'rationalisation' means that the central Coastguard Command, which receives all emergency calls, is now based in Belfast. Alerts are sent out to the smaller coastguard stations, usually manned by volunteers, in the relevant areas; lifeboats might be put to sea. The Solway Firth has RNLI stations at Silloth, Workington, St Bees, Kirkcudbright and Kippford, but there are also independent Inshore Rescue Stations at Maryport and on the Nith. Rescues are usually a cooperative process, and the lifeboat and coastguard stations have regular joint exercises in the Firth. At Kippford, a small tractor pulls the lifeboat out of what looks like the converted sitting room of a small house; at Workington, the large All-Weather Shannon-class lifeboat, the *Dorothy May White,* which is powered by water-jets, is lowered into the water on a cradle from a davit.

I drove the previous All-Weather Lifeboat (ALB), the *Sir John Fisher.* I'm still slightly horrified when I remember this. I'd been gathering information about the Solway's lifeboats for an article and had contacted Workington's coxswain, John Stobbart, to ask if we could have a chat. He called back a few days later to say they were going down to Whitehaven on a fundraising trip, and did I want to come? This was in 2016, and was part of the *Sir John Fisher*'s swan-song; she had been in service for twenty-four years and, as one of the few remaining Tyne-class ALBs, was to be retired from the RNLI's fleet. This, then, was a trip to launch the appeal for funding her replacement, the new Shannon-class ALB.

The RNLI station is on the quay outwith the dock gates so that the boat can be launched at any time. I watch as the lifeboat is hoisted from its tracks outside the boathouse and is lifted sideways on the slings, out over the water. The boat is lowered, bringing the deck level with the quay, and I and the crew step aboard. I stand by the aft hatch, trying to keep out of the way, watching the business of getting the boat ready for sea. The boat is being lowered the final few metres into the water, it's suspended over the sea –

'Ann? Come and stand here.' You don't argue with John Stobbart. He's a tall, imposing man with a gruff voice. 'You're going to steer,' he says. 'As soon as you see we're in the water – we're in it *now* – you're going to go astern . . . These two handles—' he takes my right hand, 'pull them back towards you, see here where it says "Astern". Keep the wheel where it is, see, here at zero.' Next to the wheel is a metal plate marked in degrees, zero at the bottom, increasing in steps of 10° in both directions. The engine mutters and the boat inches backwards out of the slings. I can't see clearly because my helmet keeps slipping down over my eyes.

'Now, you're going to turn, and follow the other boat out.' I have to push both throttle levers forward, gently, turn the wheel hard round to the right – and the bow swings to starboard and we circle to point out towards the sea. The 'other boat' is Workington's inflatable Class D Inshore Lifeboat (ILB), the *John F Mortimer*, with a crew of three. It's light and manoeuvrable, and one of the crew tells me, 'We use it for recovering persons stranded on rocks, because it can go in shallow water. We can even beach it if we have to.'

'Follow the other boat,' John says again. Out through the mouth of the Derwent estuary, between the piled boulders of the sea defences on the left, and North Bank on the right. The *Sir John* veers towards the sandbank.

'You're too busy talking, Ann, you need to turn ten degrees to the left.'

I stop my nervous chatter. I need to get used to the slight delay in the boat's response. The Solway is flat and grey, the wind-turbines are motionless – what must it be like to steer this boat, to make intricate manoeuvres near a broken boat, a person in the water, in huge seas? Now we're out into the channel. 'Push the levers as far ahead as you can, both together.' Our speed increases dramatically, the bow lifts, and I can no longer see what's ahead. John clicks down the two switches to the left of the wheel, which activate two hydraulically operated plates at the keel to change the trim, and the bow drops: visibility is restored. 'See that yacht there? Head inside it. Keep to the left.'

And so I steer this sturdy lifeboat, in its smart blue, red and yellow livery, on a slightly nervous and wavering course along the coast, south towards Whitehaven. There isn't time to enjoy the new perspective of the coast. Members of the crew are standing around on deck or busy in the cabin; someone is taking photographs (I see later that my smile looks like a rictus of fear). I snatch a quick glance astern at the ILB which is dashing around us, bumping through our widening wake and out into the Firth. After about ten astonishing minutes (I really am 'driving' a lifeboat!) John reclaims the wheel so that I can go below to see what is happening in the cabin. I'm allowed to take off the helmet. We soon enter the open dock gates at Whitehaven where a tent has been erected, and the press are waiting with the supporters, who are ready to encourage visitors to donate and buy a 'Keep calm and love a Shannon' mug.

*

The late George Davidson, ship's pilot for Kirkcudbright, wrote that the removal of silt from the harbour in the 1950s required 'persistent hard work with a high pressure hose and judicious moving and laying of sandbags to divert and use the flow from the main sewer'. Before the Shell-BP tankers could use their new depot more drastic measures were necessary, and in 1956 the hoses were abandoned and a dredger employed. 'The plan now was to dig out the silt, load it on lorries and dump it on low ground in the Tarff Valley. The vehicles, loaded with slopping liquid mud, splashed their way along [the town's streets] day after day. And as it dried, the mud which had spilled along the route became a very fine powdery sand, airborne with the passage of vehicles and in the lightest breeze.'[11] Dusters and Mansion House polish were doubtless in great demand.

The silting-up of estuaries and firths is a natural phenomenon – the tides and rivers carry sediment and dump it where they meet or where the currents are diminished. This is not necessarily a problem – the labile margins of the Upper Solway are shaped by this process, and the natural inhabitants can usually redistribute themselves accordingly – but where we use the edges for ports and marinas, for commerce and leisure, the clogging effect of the silt is more than an annoyance. There are several different ways of removing silt and several of them have been used (and photographed by the ship-spotters) at Workington, such as bucket-grabs or suction hoses, where the silt is picked up and stored in the dredger's hold before being dumped at the designated areas in the Firth. At Maryport Marina there is an ongoing process to stir up the silt with pressure hoses, and the suspended sediment is then flushed out by the River Ellen and the tide when the dock gates are opened. At Silloth, a dredger has recently been at work, but the usual means of clearing the outer Marshall Dock is to open the New Dock gates and let the previously contained water flush away the mud. When Annan harbour was active, a sluice between the river and the harbour could be opened – although men with shovels were also employed to labour in the mud.

The shifting of mud and sand is a means of shifting animals and algae too. Some may not survive after being deposited in deeper water, and many will be swept away to new neighbourhoods. This, and the taking-on and emptying of ballast by ships – either hard ballast before about the 1860s, or water thereafter – is a means of spreading marine organisms like the Fundy Bay mudshrimps around the world. Where the organisms attach to solid surfaces, the hulls of ships are vectors too. In the same way that 'non-native' plants like Himalayan balsam and rhododendrons have spread and found new homes in parts of these islands, so have non-native animals and seaweeds colonised our maritime edges. They even have a name: Marine INNS – marine invasive non-native species.

Marine INNS are in general fouling organisms that are highly adaptable and so can multiply under a range of environmental conditions; they include various algae and animals like ascidians (tunicates or sea squirts), sponges, barnacles and bryozoans (moss animals). They are readily transported around the globe attached to ships' hulls or as planktonic larvae in ballast water, and so ports and harbours – with their many hard substrata such as pontoons, mooring-ropes and buoys – are ideal sites for colonisation and spread. As happens with most newly introduced species, in the absence of the regulatory effects of their usual competitors or predators, they often out-compete local species. An example is the slipper limpet, which has the all-too-illustrative name of *Crepidula fornicata*, because it forms vertical piles of individuals that change sex as they grow older; it was introduced to Essex in the late nineteenth century with American oysters *Crassostrea virginica* from the north-east of America, and it may also have been transported both on ships' hulls and, in its larval stage, in ballast water. Since 1934 it has been found in silt brought into the UK with Dutch oysters, and it is now abundant in the whole of the North Sea on oyster and mussel beds – like them, it is a filter-feeder, but it out-competes them for food. *Crepidula* has not been found in the North-West, but the Solway Firth Partnership (SFP) has been monitoring the presence of other Marine INNS in the Firth, with surveys and by hanging 'settlement panels' in various harbours and marinas to see what attaches with time.[12] Hayden Hurst, who worked with Cumbria Wildlife Trust, found dense settlements of eleven different INNS (as well as native species) during surveys at Fleetwood on the Lancashire coast, and SFP say the panels at Stranraer in Loch Ryan are also heavily coated. Both these ports, to the south and north respectively of the Firth's entrance, host a large amount of marine traffic, from ferries to trawlers to tankers. Hayden also found 'superabundant' colonies of the reef-building trumpet tubeworm, *Ficopomatus enigmaticus*, at Whitehaven – and one specimen on a panel at Maryport, apparently where a boat from Whitehaven had been moored (this polychaete worm had not been seen there subsequently). The tubeworm colonies 'led to major fouling of pontoons and yachts in the marina', being ten centimetres thick in places – this would 'significantly impede vessel movement due to drag'.[13]

Other INNS found commonly or only occasionally in the Firth's harbours and marinas have included the leathery sea squirt, *Styela clava*, which can be as much as twenty centimetres tall; the orange-tipped sea squirt, *Corella eumyota*, and Darwin's barnacle, *Austrominius modestus*. Some of these are a nuisance in fouling marine structures and the bottoms of boats; some out-compete local species. Study of the settlement panels and rapid surveys of pontoons also found, of course, many native species; I was happy that Hayden had recovered small numbers of native *Corophium* species from panels at Whitehaven and at Fleetwood.

In general the number of species of INNS and the size of their populations along the Solway coasts is much less than elsewhere around the British coasts, perhaps because the Irish Sea's coasts act as a buffer, and perhaps because the salinity of some of the harbours next to rivers, like Maryport, is very variable. INNS have to be reported, and there are national and European regulations that cover this, as explained in the SFP's reports. There are also well-publicised 'biosecurity' techniques for reducing introduction and spread, including 'check, clean, dry' of marine equipment, and regulations for the disposal of ballast water.

Most INNS, though, are here to stay. Fifteen years ago at Campfield saltmarsh Norman Holton showed me *Spartina*, the common cordgrass, originally an invader from America, which was colonising the mudflat and lower zone of the marsh; it is along the merses around Caerlaverock too. Sargasso seaweed or wireweed, *Sargassum muticum*, grows at Loch Ryan just north of the Firth and is heading northwards up the west coast, but is apparently not yet growing in the Solway. Where it does grow it forms dense mats that could get tangled in ships' propellors and stifle other growth, but in its natural habitat it also provides food and shelter for a wide variety of organisms, from shrimps to turtles. Recent evidence from tagging studies has confirmed the long-held supposition that the Sargasso Sea is indeed the place to which European eels migrate to spawn.

The quick method for distinguishing wireweed from our native wracks and kelps is to hold its ends and spread your arms apart; the main frond is like a washing-line from which the lateral branches dangle, knobbled with air-sacs like glossy beads.

Sargassum muticum is sometimes jumbled amongst other seaweeds and flotsam along the tideline near Allonby Bay. It is a robust and assertive seaweed or macroalga and this makes Anna Atkins' cyanotype of *Sargassum plumosum* even more impressive. Collecting seaweed and displaying it in albums was one of the 'sea-side pleasures' that entertained the Victorian middle classes. Ladies were not expected to be scientific in their collecting but, rather, to arrange their seaweeds in an aesthetically pleasing manner; the making and displaying of scrapbooks was a parlour pursuit, art rather than science. However, some ladies took the study of seaweeds more seriously. There was Amelia Griffiths (1768–1858) at Torquay, who became very knowledgeable about the seaweeds of the Cornwall, Dorset and Devon coasts, and helped the Irish botanist William Henry Harvey (1811–66). He produced the handbook of *British Marine Algae*, followed later by the three-volume *Phycologia Britannica*, illustrated with coloured plates and published in 1846, in which he references Amelia Griffiths (sometimes as the 'indefatigable Mrs Griffiths') nearly sixty times.[14] He includes some details of *Sargassum bacciferum*, which he acknowledges 'has no claims to be admitted to the British Flora' although it is occasionally thrown up along northern shores; his illustration is of a 'specimen picked up at sea, from the great floating bank of gulf-weed which extends at the westward of the Azores'.

While Harvey was working on the accurate representation of British seaweeds through coloured engravings, the Fellows of the Royal Society were learning about other methods of capturing images – on paper or metal plates that had been made sensitive to light. William Henry Fox Talbot (1800–77) had been developing a method of recording images as negatives on high-quality paper sensitised with a solution of silver iodide, which became known as the 'Talbotype' or calotype process. Astronomer and chemist John Herschel (1792–1871) had been testing the effect of the spectrum of light in changing the colours of plant extracts. He began to focus on the underlying chemistry and the best methods of producing – and fixing – the colour change of the pigments on exposure to sunlight, and he used this to produce negative images of objects placed upon the sensitised paper: a process he called *cyanotype*. Subsequently, he abandoned plant extracts, with their messy cocktail of constituents, and started to sensitise paper with a mixture of iron salts, ferric ammonium citrate and potassium ferrocyanate. Exposure to sunlight produced a background of 'beautiful and pure celestial blue' – Prussian Blue – around the object that was being recorded.

At roughly the same time as Harvey was working on his *Phycologia Britannica*, Anna Atkins was producing the several parts of her

Photographs of British Algae: Cyanotype Impressions, each of which contained around 400 images.[15] It is not known how many copies she made but fewer than twenty still survive, all slightly different. Anna Atkins (1799–1871) had been fortunate in that she had received a broad education, especially in 'natural philosophy', and that her father John Children and her husband, John Atkins, mixed with scientists and Fellows of the Royal Society. They met socially too with Fox Talbot and Herschel, and so Anna learnt about calotypes and cyanotypes, the new techniques of photography, from these friends. She began recording the variety of British seaweeds using Herschel's photographic technique of 'blueprinting' or cyanotyping, placing pressed algae under glass on paper that had been sensitised with a mixture of soluble iron salts, and exposing them to the sun. After the paper had been washed in water, the background colour deepened to the uniform 'celestial blue', leaving the detailed silhouettes of the algae white; drying the paper fixed the colour and the image. The algae are so clearly displayed and their outlines so sharp that they can easily be identified as to species. This is all the more impressive because (as I have myself discovered) thicker algae like *Sargassum* and *Fucus* require a longer exposure during which time the angle of the sun moves ever so slightly, potentially causing fuzziness at the edges. *Sargassum plumosum* is (probably) the same as *Carpophyllum plumosum*, which still belongs to the same taxonomic family, Sargassaceae. Anna Atkins made an early photographic representation of a Marine INNS.

One of the reasons why film-maker Julia Parks and I met at Parton beach was because, unlike many of the English beaches along the Upper Solway, it is not soft-edged but rocky, and has a good variety of macroalgae exposed at low tide. The delicate reds and greens are especially beautiful when placed in fresh water and floated onto card: large polysaccharides like alginate and carrageenan are released from their cells, swelling as they hydrate, and glue the fronds onto the background as the specimen dries. I still have some of these cards, made during seaweed identification classes when I was a student, and the colours and shapes of the macroalgae are as bright and fresh as the day they were collected. Less scientifically (and with my rueful nod towards those Victorian ladies), my grandchildren use the same technique to make seaweed pictures. The cyanotypes, though, have an absolute requirement for sunny days, and although they are fun to do, they are also challenging. Julia has been using seaweed – the bladderwrack, *Fucus vesiculosus* – for a different reason: specifically to develop her 16mm films. Having discovered that the best technique is to use seaweed mixed with vitamin C and washing soda, Julia pours boiling water over the mix and leaves it to 'brew' for twenty-four hours; the film is developed in the cooled extract. Her first experiments with the technique were carried out at Allonby, and BBC's *Countryfile* filmed her carrying out the technique with Matt Baker's help on Roa Island off Barrow.[16] The slightly grainy, softly muted contrast of the black-and-white film perfectly captures the subdued light along the coast.

Should one of the various proposals for tidal power lagoons in the Solway Firth ever reach acceptance, the amount and variety of shipping in the Firth and around the ports will increase dramatically, with surveying rigs, barges carrying cement and turbines and rock armour (one suggestion is that a new jetty should be built for the quarry near Creetown to allow access to its granite), dredgers and a multitude of support vessels. For several years there would be plenty of activity in the Solway for Julia to film. Two of the tidal power plans also incorporate seaweed farms within their lagoons.

4

Marshes and merses

At first light on a Sunday morning in late September, Norman Holton sat on the edge of Campfield saltmarsh on the English side of the Firth. On the Scottish side the starlings were gathering in great wheeling clouds, and as usual there were several sparrowhawks flying above them, attempting to pick off a few for breakfast. This time, though, the starling-cloud spiralled round and round, and the mass of birds coalesced and flew across the Firth. Norman estimated that there were two million birds: 'They were flying low, about ten feet above the water, coming straight towards me. I couldn't see the ends of it, from Cardurnock to beyond Herdhill Point, the flock was so wide. It must have taken ten minutes to pass over – it lifted slightly to pass over the marsh, flying right over my head. The noise! And the wind of their wings, the draught! I was absolutely plastered in crap. But it was fantastic – the hairs on the back of my neck are standing up just telling you about it!'

Until his death in 2016, Norman Holton was Senior Sites Manager of the RSPB's Cumbria Coast Reserves and not one to exaggerate about bird numbers. The Campfield Reserve, based at North Plain Farm between Anthorn and Bowness-on-Solway, has two miles of coastline edged with saltmarsh and extensive mudflats, as well as a large expanse inland of arable land and raised bog, and it is host to thousands, sometimes tens of thousands, of wildfowl and wading birds. But on the day I met Norman, more than fifteen years ago, the tide was well out and the bird-flocks had dispersed to feed across the miles of glistening muddy sand. Our wellies made perfect prints in the overlying layer of mud as we, too, stepped down from the cropped grass of the saltmarsh and walked out into the estuary. There was activity all around us: small shore crabs scuttling; patches of tiny black flecks that were snails, *Hydrobia*; the surfaces of shallow pools suddenly churned by the skittering of minute fish and shrimps; muddy coils ejected by lugworms; minute holes made by burrowing mudshrimps; and the empty shells of cockles and pink tellins that, when alive, had burrowed in the shore near the low-tide mark. Even though the Scottish coast was only a couple of miles away across the Firth and houses and cars were startlingly visible, there was space and emptiness – and silence. Sheep were grazing the marsh in the distance, the sun was a pale disc in the misty air, and for a while, until the tide turned and the birds returned, there was peace: a privilege.

At such times, I realise how important it is to be with someone who knows and cares deeply enough about a place to be silent: words are unnecessary as we watch, and listen, and see.

*

Saltmarshes (or 'merses' in Scotland) have a similar basic structure whether they border estuaries or firths. On that warm August day when I met up with Norman, he led me across the Campfield Marsh, stepping over deep potholes with overhanging edges (known locally as 'dubs' – pools or water-filled holes), jumping across slippery-sided creeks, and following the cattle-trails that divert around the meanders and oxbows of the estuary's probing fingers. The profile of a saltmarsh is always changing, a balance between water and salt-tolerant plants, but despite the mutability of its lower reaches, if you look carefully at the marsh and its surrounds you can see that there is a logic to the structure, because it is a record of time, as well as the tides.

The single-track road that follows the coast is built on a raised beach, and the marsh steps down from it in three tiers. At the top there are

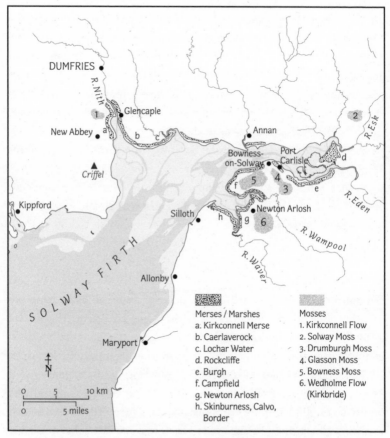

Merses, marshes and mosses

gorse thickets and tough creeping couch-grass, thistles, clumps of pink-flowered rest-harrow, and low purple asters; damp hollows are filled with the rush, *Juncus*, and sedge. Walk towards the sea and you step down a small 'cliff' onto the next tier, where the salt-tolerant grass, *Pulcinellia maritima*, is like shiny wire, and there are low broad-leaved plants like silverleaf. Here too are jagged creeks, where the water has felt for weakness and burrowed into the body of the marsh; waders have left their sharp-toed footprints and probe-marks where they have hunted for snails and polychaete worms along the muddy bottoms.

The marsh's vegetated cushions, speckled with pink thrift, taper down towards the sand and there, at the edge and marching outwards towards the sea, are the 'pioneer species': solitary, upright, their limbs pointing defiantly upwards, they appear intrepid and surreal. I have a distinct impression that they will advance a few centimetres, triffid-like, the instant I look away. Some have gathered a little sand around themselves to form an embryo island, some of the islands have accumulated a tuft or two of grass, each island will consolidate and grow and the marsh will spread outwards. The plants are samphire or glasswort, *Salicornia*, pale green and fleshy. Norman tells me he used to fry them in butter and eat them when he was an impoverished RSPB worker on the Wash, and he picks off a piece for me to try. It is juicy and salty and delicious. I am an instant convert. On the seaward side of a low green island is a mat of a surprisingly spiky plant, *Spartina anglica*. *Spartina*, too, is a pioneer and in more than one sense, for the genus is an invader from America.[1] 'It's an absolute pain,' Norman tells me. 'Once it gets a foothold it spreads and spreads.' The seeds come in on the rising tide, and get deposited as the tide goes out; they germinate and grow and trap silt, raising the level of the marsh.

Samphire is much in evidence on the Scottish side of the Firth too, where the Caerlaverock merse fringes the eastern side of the River Nith, but this time I am visiting on a chilly day in March 2019 and the plants look grey and wizened, worn down by the winter; there are still traces of snow on Criffel's northern slope. Adam Murphy, Nature Reserve Officer for Scottish Natural Heritage's Caerlaverock

National Nature Reserve[2] had responded enthusiastically, immediately suggesting an 'expedition', when I had asked him about the presence of mudshrimps on the reserve's mudflats, and after I meet with him and another Nature Reserve Officer, Andy Over, in the car park we drive over to Scar Point on the east bank of the Nith. Scar Point is also known as Fishermen's Bush (it's a favoured spot for fishing when the tide flows in) or Phyllis's: Phyllis Laurie, who died in 1942 at the age of seventy-two, lived nearby and according to the Solway Firth Partnership's little booklet *Tide Islands and Shifting Sands*[3] was a 'character' who was 'feared by local children'. She owned a horse and cart, and the story has it that if she became incapable of driving after visiting the pub on a Saturday night, the horse always knew its way home. Phyllis might be pleased that the Point has been given her name – but perhaps unimpressed that her name also distinguishes a nearby creek across the mud.

Adam strides out onto the merse, while Andy – recently arrived at Caerlaverock from ten years with SNH on Harris – offers to carry the spade that I have brought and kindly holds out a hand where creek-jumping requires longer legs than mine. As we walk across the merse Adam points out the partly exposed gnarly rhizomes of sea-aster, the fleshy spikes of arrowgrass and the shiny leaves of two species of scurvy-grass, amongst the thin film of sediment on the short 'turf' of salt-tolerant *Pulcinellia*. When poet Norman Nicholson wrote about the saltmarshes on the Solway's English coast in the 1940s, he was writing during a time when iron ore smelting and steel making around Workington were important parts of the West Cumberland economy. He was attracted to the marshes even when 'as around Workington and Maryport, they are soiled and smeared by smoke and scum from the iron and steel works'.[4] The residues of industry affected even the white flowers of scurvy-grass – 'Many people despise this plant, perhaps because . . . it gets clogged with coal-dust as in Workington Harbour, or with red ore-dust in the iron country.' Those industries are long gone, and in April the scurvy-grass flowers are 'a welcome mass of white', and the surfaces of the saltmarshes unsullied. Adam told me that sailors used to eat it as a source of vitamin C. Norman Nicholson notes, 'It must have been nearly as unappetising as boiled cabbage.'

We find otter spraint, glittering with fish scales, and discuss whether the folklore that it smells of violets is true (it does not, but it is not unpleasant); and wind-blown mermaids' purses, these ones the square matt-black egg-cases of the thornback rays that are common in the Firth. The north-westerly wind is ferocious and cold, blowing our words away and tearing the pages of my notebook. Five pinkfooted geese come beating across the mudflats, heading slowly into the wind, their wingtips almost touching the mud. The edge of the merse is ragged and unfocused, a perfect example of the estuary's give-and-take. Little troops of samphire had boldly marched out during the previous summer, and here and there had gathered new ground before losing heart and shrivelling. Further out are scattered islets, new 'lumps of merse' as Adam calls them, now stabilised by thin patches of grass; he has watched how the merse has pushed outwards in just two seasons. Now, at a low spring tide, the estuary's width is obvious and Adam explains how the river channel has shifted over to the western side, leaving great sloping mudbanks here on the east. But deep crevasses are carved through the central whaleback of mud, as though the river is trying to break back through again to nibble at the newly accreted merse. At Campfield, I had seen metre-high towers at the seaward side of the marsh, capped with vegetation but with steep, bare sides. Some of the towers and hummocks had collapsed and were being smoothed and redistributed by the sea. Here, too, at Caerlaverock the tides had picked away at the lower tier, shifting the mat of vegetation and exposing compacted mud and a sharp 'cliff-edge'. Columns of merse had been isolated. 'We call them merse-bergs,' Adam says. 'If you're out here on a quiet day, sometimes you hear a rumble, a bit like thunder – and then you realise that one of the bergs has collapsed.'

Growth-rings in trees, fish scales, stromatolites and oyster shells: the laminations are the sequence of responses of living organisms – plants and bacteria and animals – to diurnal and seasonal changes in their environment. The bodies of the bergs and towers of the saltmarsh are not alive, but where they have been sliced open by the tides their inner structure too is exposed, striated with subtly different colours and textures, bands of jagged white shell fragments and rounded pebbles between the layers of silt, a record of particles that have been swirled into their environment throughout the years. The marshes and merses erode and accrete, the removal and deposition of sediment changing with the weather and the tides, the seasons and the years, and on geological timescales.

*

There is something else that is ancient living on the merse at Caer-laverock. *Triops*, the tadpole shrimp, is unrelated to *Corophium*, the mudshrimp, and the two species are unlikely ever to meet – but *Triops*, a star of the Solway, is not to be ignored and must be included here for completeness. The modern story of *Triops* is one of discovery, loss and rediscovery.[5] This is an animal that has scarcely changed since the Triassic period – its fossils date back 200 million years. It is a freshwater crustacean, belonging to the Order Notostraca, and looks very similar to (but is unrelated to) a small horseshoe crab as it trundles around on the bottom of a pond: its head and thorax are covered by a carapace, like a shield, so that from above its legs and mouthparts are scarcely visible.

Was it always the case that *Triops* lived, fed and bred in freshwater pools? This is a dangerous life-strategy yet despite, or because of, this danger the animals evolved a means of surviving when the pools dried out. Today, they are found only in ephemeral pools in the New Forest and on the merse and wet pastures around the Wildfowl and Wetlands Trust (WWT) Caerlaverock Reserve. They can live for two to three months, or until the tide inundates the merse, and when the pools dry out under the hot summer sun the adults die and disappear. But in the drying sediment their eggs live on, 'switched off' in a state of diapause, or suspended development. When the rain comes and the pond re-wets, the eggs are stimulated to hatch – and the larvae feed and grow very quickly, so that adults are ready to lay eggs within as little as two to three weeks from hatching. As Dr Larry Niven, Principal Species Research Officer at the reserve, says, 'They can flash in and out of existence – it's just luck whether you find them. It needs an inquisitive person . . . in the right place at the right time.'

It was more than 100 years ago that *Triops* (then known as *Apus*) was found on the merse by F. Balfour-Browne. Forty years later, in 1948, in a paper to the scientific journal *Nature*, he wrote (or narrated – the style of scientific writing was much more like storytelling than it is now):

In September 1907 I discovered two shallow grassy pools on the Preston sea merse, near Southwick, Kirkcudbrightshire, in which Apus was present. In one of these it was so abundant that when I raised my eleven-inch ring net out of the water it was half full of specimens, mostly full-grown. I searched many other pools in the same area but without finding it and, returning to the same pools a few days later, I found the edges covered with the shells and very few specimens left in the water. The gulls had discovered this mass of food and had destroyed most of the Apus. I have visited the area many times during the last forty years but not until this month, working the merse near the mouth of the Southwick burn, have I again seen Apus. My son found three specimens in a pool which then yielded us about thirty or more, and several other pools near the first produced small numbers, mostly immature.[6]

These pools were subsequently lost as the merse was eroded, but Larry Niven found the tadpole shrimps again in 2004, nearly twenty kilometres away, and since then three more sites have been discovered. It is thought that eggs are spread in several ways – through the guts of animals, or on the feet of cattle, deer, or geese, 'Anything traipsing around, really.' This includes the tyres of tractors and quad bikes, and because the tadpole shrimps are hermaphrodite, 'It just needs one egg,' Larry says, 'and after it hatches and survives, you could then get hundreds.' Sometimes the shrimps themselves have been found, at other times it has been their eggs, discovered in washed and filtered sediment samples. Now, though, there's a new molecular diagnostic technique available, to test environmental samples for a range of 'eDNA' – fragments of genes that are species-characteristic and which may have been left 'lying around' in the environment. Graham Sellers from Hull University has recently developed this 'DNA bar-coding' technique for *Triops* and, working with Larry, has established the existence of the shrimp in six ponds at Caerlaverock.[7] Perhaps this technique could now be used for pools on the other Solway saltmarshes. It would be exciting, and a great relief, to find that the story of the rare and elusive *Triops* could be expanded to a few more chapters.

*

After leaving the SNH Caerlaverock Reserve, I drove eastwards along the back road parallel to the Firth, to view the expanding merse by Lochar Water. The narrow road was long and temptingly straight, but winter-bred potholes were a stronger deterrent than any speed cameras. Geese, both pinkfooted and the delicate black-and-white barnacle, were grazing the low-lying fields. The entire Svalbard population of barnacle geese arrives in the Solway region in October and may stay as late as May, fattening up before flying north again to breed. Pinkfooted geese over-winter on the Solway, the Wash, the Ribble and eastern Scotland, having flown here from Iceland, Spitsbergen and Greenland. Some of these travellers make the journey in a single day, and in October the chattering discourse of geese overhead – *wink-wink-wink-wink* – causes people to look up, search the sky for the skein, and smile. 'They're back!' Break-away groups, dissidents, a change in the leader of the 'V' – the flight-patterns constantly shift and re-form. Birdwatchers and recorders are busy on Twitter and Facebook throughout the winter and spring, posting counts: 'Three thousand pinks at Calvo', 'A leucistic [white variant] barnie at Campfield'. On the South Solway, a 'co-ordinated dawn roost count' recorded 34,500 pinkfoots in February

2019, and an estimated 7,000 barnacle geese were grazing the marsh at Moricambe Bay in September 2018. There is no competition with four-footed herbivores – the cattle are not put onto the marshes until May – but the Secretary of the Newton Arlosh Marsh Committee told me, 'If the marsh grass doesn't look good the geese'll come into the barley fields, and I usually let the farmer know so he can chase them off!' David Campbell of the Scottish Solway Wildfowlers Association (see also Chapter 9) says that 'without a shadow of a doubt' the number of pinkfooted geese has increased enormously in the past decade or so, and that their behaviour has changed too, so that they often roost on the sands and merse; but he is especially intrigued by their grazing behaviour when they are feeding on grass: 'There's always been differences between the farms – the grass looks the same, but there are some farms [near Caerlaverock] that always have geese on the grass, but the one next to it, never. Why is that, and how do the birds know? Do they pass the information down from year to year?'

At Brow, a short track beside the burn led through a gate onto the merse, where a pole carried the metal silhouette of a flying goose, although it seemed to have crash-landed vertically on its tail, and a notice told of the Scottish Solway Wildfowler Association's interest in this area. I walked towards the mudflats of the River Lochar and suddenly hundreds of periscope-like heads were raised and swivelled, as a somnolent flock of geese saw me appearing like 'an upright spelk on their shelf' against the flatness, as poet Tom Pickard wrote of his own presence on the shore in *Winter Migrants*.[8] The water was far away on this low spring tide, and even though the merse was not such a wide intermediary in his time, I imagined poor Rabbie Burns waiting for the tide to return in July 1796. He was very unwell – recent interpretations are that he was suffering from subacute bacterial endocarditis – and had been prescribed a 'cure': to drink from the chalybeate spring at Brow, its waters brown and murky with iron salts, and then to wade out, up to his armpits and fully clothed, to bathe in the Solway Firth. He subjected himself to this miserable treatment a couple of times but sadly, and unsurprisingly, his condition deteriorated very rapidly and he died a few days later on 21 July in Dumfries. Every year a commemoration service is held at Brow Well to mark the poet's death, and in 2016 the well and its surrounds were improved with the help of Solway Firth Partnership and others. Now the cistern is surrounded by fine red sandstone paving engraved with a verse of Burns' *A Prayer, in the Prospect of Death*: 'O Thou unknown, Almighty Cause Of all my hope and fear! In whose dread Presence, ere an hour, Perhaps I must appear!' Above a spout a notice requests 'Do not drink the water'. It is hard to imagine anyone wanting even to dip a finger in the brown scummy water that was dribbling from the pipe. Brow Well does not exude cheerfulness.

Burns must have come to know the Solway and its tributaries well during his job as an excise man for the port of Dumfries, but although John Young has researched Burns' works for mentions of the 'natural world',[9] the Solway appears only once, in a dismal connection with the River Nith:

> Peg Nicholson was a good bay mare,
> And ance she bore a priest;
> But now she's floating down the Nith,
> For Solway fish a feast.[10]

*

Summer 2017, Rockcliffe Marsh. Two lines of hoofprints, large and small, drop down from the saltmarsh and meander across the sand towards the low-tide mark, then loop back landwards. The heifers are no longer in sight; indeed only a few of the several hundred head of cattle out on the Marsh are visible, as black- or brown-and-white specks, so vast is Rockcliffe Marsh. Wind hisses across the drying sand of the empty foreshore; an oystercatcher trills; a heron calls harshly, once, as it flaps heavily across the estuary to Scotland.

Rockcliffe Marsh is surrounded on three sides by water: it lies at the head of the Firth, bounded by the River Eden on the south and a loop of the River Esk on the north. At that time, in 2017, Imogen Rutter was Cumbria Wildlife Trust's summer warden, employed to monitor the numbers and species of breeding birds on the Marsh. She is my guide on a quiet but overcast May morning, and we walk out along the high embankment that has been built to protect the fields and byres on the landward side, then drop down onto the Marsh. Within a half-hour we are far out amongst the cropped turf and creeks. It would be easy to lose our bearings, without the distant bank to orientate ourselves. There are other markers too, less easy to see on the Marsh's slightly undulating surface: wooden stakes mark the few bridges across the creeks, and a dotted line of white posts shows a route for wildfowlers, indicating where they may cross the Marsh but may not shoot. By one of the bridges we find the scattered remnants of a gull. We poke around looking for the leg-ring and find it on a dismembered bright-orange leg: it had been ringed in Norway at Skagerrak Museum, and had died in Cumbria.

Neilson, in his 1899 *Annals of the Solway*, comparing maps from 1590 to 1895, found a 'very great change', for the marsh had extended two miles west out into the Firth in that time.[11] Since then it has grown and grown, reaching out between the two great rivers. It is one of the largest saltmarshes in Britain, at about 1,100 hectares, but its margins are constantly changing; one year it added twenty-six hectares to its width. This, and the fact that it is farmed as well as having multiple layers of protected status, makes it an intriguing and special place.

The Marsh and its considerable foreshore are owned by Castletown Estate, and when I contacted the owner, Giles Mounsey-Heysham, in mid-July of 2017, he immediately suggested we meet and he would tell me more about it. As we chatted, we looked at photos and plans, and maps spread out on the table in the estate office. Measurements of the Marsh's perimeter have been made since 2001, using GPS and a quad bike; more recently the Environment Agency's LiDAR maps of elevations across the Marsh are being used to inform work on water retention. Giles' enthusiasm for the Marsh was obvious and he was very keen to take me out and show me its many facets, so he suggested that he take me out on the back of his quad bike. It was well worth the couple of hours of discomfort, occasional rapid elevation from my seat (and the application of Savlon when I returned home) to travel across and around the margins of the Marsh.

Many of the creeks were small or dry, but all have names. We made a large, looping diversion to avoid the wide inlets of Stony and Yellow Creeks and later, at Near Gulf, Giles told me to climb off and wait while he drove the quad down the muddy bank into water that churned black under the wheels and up the glossy brown incline on the opposite side. It seemed that a crossing was possible – he returned to collect me. Judging by the sheets of muddy water that sprayed into my face, the trick for not getting stuck is to drive fast through the water. To the south on our left was a raised carpet of gold, where the glittering dead stalks of thrift had formed a glorious pink cushion not long before. Norman Nicholson, in his 1949 book *Cumberland and Westmorland*, mentions thrift, 'the flower of the threepenny bit'. Few people would understand that allusion these days but in a little pot retrieved from my parents' house after they died, and amongst the 'lucky' silver sixpences (to be covered in tin-foil and buried in the Christmas pudding) I was thrilled to discover a 1937 twelve-sided 'Three Pence' coin, with George VI's head on one side and the visual pun on the reverse.

This huge new area of saltmarsh to the south, consolidated by grass and thrift, had developed in the past six years: south-east, near the mouth of the River Eden, an island had recently grown and was already hazed with green; to the north, at the mouth of the River Sark near Gretna, another embryonic saltmarsh was developing. From Sarkfoot Point we could see the distant stream of lorries and cars grumbling along the motorway to the east, where Metal Bridge crosses the River Esk. We dropped down onto the foreshore, which stretches way out into the Firth at low tide; aerial views reveal hectare upon hectare of mud and sand. The mud was pocked with the holes of mudshrimps, their feeding-marks radiating from the openings; the feet of tiny *Hydrobia* snails had ploughed furrows where they foraged. Out on the firm sand the exposed bed of the Firth was patterned in the colours of desert camouflage, but would be covered with metres-deep water when the tide flowed in.

Back in May, Imogen and I had picked our way to the Marsh's edge on foot, across the uneven sward and around the creeks, larks filling the air-columns of their territories with song, occasional lapwings *phweet*ing and diving around us. Then suddenly, a sheet of birds had risen up in the distance, wings beating heavily at take-off. Flighting, the flock came towards us – hundreds of barnacle geese, flying low over our heads, talking to each other, perhaps grumbling at the disturbance, and heading across the Eden to Burgh Marsh. And then another black-and-white sheet rose, and then another, stirring the air with their wingbeats. My skin prickled as geese flew over and around us, changing the Marsh's character, inhabiting the air completely with their bodies and their sound. But in a few minutes they were gone. Within a week they would probably have departed until the autumn, having built up their strength by grazing on Solway grass for their long flight back to Svalbard, and their short breeding season.

Rockcliffe Marsh provides grazing for farmed mammals and for geese, but it's not just a vast expanse of pasture. It also has many national and international conservation designations, an alphabet soup of acronyms.[12] Because the marsh is so comprehensively protected under national, European and international laws, its management is overseen by Natural England (NE) as the government's statutory body. I met NE's Dr Bart Donato on a sunny morning outside a café near Kendal, and during our amusing and long conversation, when his enthusiasm for Rockcliffe spilled over into making models of creeks and levées with pink and blue play dough, he explained that it is also managed as a saltmarsh, to create a mosaic of wet ground and creeks and open water amongst the vegetation, 'a Swiss-cheese effect of water-bodies and big sheets of water'.

The Solway Firth has famously sediment-laden tides, the sea turning the colour of milk chocolate during strong winds. When the incoming tide reaches the head of the Firth at Rockcliffe and meets the outflowing fresh water of the rivers, it deposits its load of silt. Moreover, during storms, sediment also washes down the rivers 'from everybody's fields', as Bart said, and much of this gets trapped upstream of the neck of the Firth at Bowness, so that above-average amounts of riverine sediment are deposited too. The Marsh grows outwards, but also *upwards*; this upward growth is caused by 'topping tides', the high spring tides that happen when the moon and sun are in alignment and their gravitational pull is greatest. Then, Bart said, the water, 'brown with muck', creeps in through the creeks and their overspills onto the Marsh. The vegetation creates at its base a layer of still water, a 'stationary layer' from which the muddy sediment precipitates out. On a big tide there's a relatively long period of slack water at the head of the Firth, which means a longer period for sedimentation to happen; as much as one centimetre of silt might then be deposited within just a few tidal cycles amongst the grass and herbiage. Near the elbow of the River Esk, Giles jumped off the quad and showed me a line of fence posts that was gradually getting buried, and raising his hand to show their height, he indicated they were nearly one-third taller when knocked in.

There may be hundreds of hectares of useful grazing on Rockcliffe, but there are also topographical problems, not only of creeks and sticky mud, but of river banks too. Stock are the economic lifeblood of the Castletown Estate. When Giles took over the estate, there were 1,000 head of cattle, 'with one man on horseback to keep an eye on them . . . but then the rules for the stocking rate changed and farmers were no longer interested. So we started putting our own cattle on – we now have quite a big beef enterprise, 800 cattle, of which 500 are our own.' There are also 2,500 sheep in the summer, about 800 ewes – mules, Texels and Romneys – and their lambs, although they were not out during our quad bike expedition. We saw them later in the steading beyond the embankment, being dosed and checked amid a cacophony of noise as they milled around in the pen, watched by muddy border collies. Earlier, on Eskside, about fifty gipsy horses, black-and-white and brown-and-white, raised their heads to stare, then galloped away, flanked by their foals, whinnying and kicking up their heels.

Giles told me stories of rescuing cattle mired in a muddy creek by the Eden with the tide coming in. The fire brigade, the coastguard and local helpers all worked hard to get them out: 'The fire brigade used their pressure hose to act like a lance and wash the quicksand away from around the animals' feet.' Since then the proven technique for rescuing mired animals is to bring out a quad bike and trailer with a pressure-washer and tank of water.

Brian Blake, in his 1955 book *The Solway Firth*, talks about the creeks: 'The tide rises and you do not see it because it is in the gulleys, and then in a short while the marshes seem to become lakes or fiords or even, if it is stormy, the open sea.' I too have watched the tide creeping in silently, frothy-toed, the brown water rising in the creeks and spilling over. Norman Holton told me about a day when he was bird-watching on Rockcliffe Marsh and the tide came in fast, quickly flooding onto the grass. He grabbed his tripod and telescope and hurried back towards his car, trying to watch for hazards in the turbid water – but fell into a deep creek. 'It was February, freezing cold,' he said, 'it took me ages to get to the car and I had no dry clothes.' 'So what did you do?' 'Put the heater on high, stripped off and drove home in my underpants!'

*

Skinburness is at the root of the Grune peninsula, on the south-west corner of Moricambe Bay and adjoining the Skinburness and Calvo saltmarshes. The River Waver runs close to the eastern edge of the Grune, and it is here that archaeologists have found traces of a medieval port, for this was where Edward I had his naval base during his protracted wars with the Scots. It was reported that there were more than fifty ships in the Solway in 1300, some summoned from as far away as the Cinque Ports, as well as more local craft from Whitehaven, Workington and Allonby. There were fighting ships and ships carrying victuals for the army; stores were accumulated at Skinburness. On 12 February 1301, King Edward I granted a charter making Skinburness a port and a borough for the Abbey of Holme Cultram, and in August of the same year Bishop John de Halton at Bridekirk authorised the abbey to build 'a chapel or church' at the port; the right to hold a fair or market was also granted. But this was shortlived. Grainger and Hollingwood, surveying the records relating to the abbey, write:

> For in 1305, we find thus mentioned in the parliament records; 'At the petition of the abbot requesting that whereas he had paid a fine of 100 marks to the king for a fair and market to be had in Skinburnese, and now that town together with the way leading to it is carried away by the sea, the king would grant that he may have such fair and market at his town of Kirkeby Johan [i.e. near the new Church of St John, Newton Arlosh] instead of the other place aforesaid, and that his charter upon this may be renewed; it is answered, Let the first charter be annulled, and then let him have a like charter in the place as he desireth.'[13]

Some time between August 1301 and April 1304 there was a mighty storm and the port and part of the hamlet of Skinburness was washed away. A new port, with the right to a fair and market, was chartered on the east side of the bay at Newton Arlosh. As you drive east from Skinburness towards Newton Arlosh, for some of the way the road runs between the marsh and a high grassy bank, the medieval sea dyke, built to protect against further inundation. Cattle wander along the road, undeterred by traffic, turning their heads to stare disdainfully before budging an inch or two.

It is no longer obvious where the new port was sited, for the ragged edge of Newton Arlosh Marsh stretches far out into Moricambe Bay. The new hamlet was built on land that is slightly raised above the salt-marsh. The village is strung out along the road, the Church of St John the Baptist set back on the side nearest the marsh. It's a tough-looking little church, with thick walls and a pele tower. It has many similar features to the other fortified churches of the northern Solway Plain, such as St Michael's at Burgh-by-Sands. They were to provide an easily defended building into which the locals – and their animals – could retreat when an attack was imminent: narrow entrances, strong doors, windows high above the ground, thick-walled towers with rooms that could only be reached by a staircase wide enough for a single person, arrow slits and crenellated battlements.[14]

After the dissolution of Holme Cultram Abbey in the sixteenth century, the church fell into serious disrepair. As John Curwen wrote in 1913 (in a paper that 'was read on site'), 'Under date 1580 we read "The chapel of Newton Arlosh did decay; the door stood open, sheep lay in it. About fifteen years since the roof fell down and the lead was taken away by some of the tenants and converted into salt pans".' Curwen continues, ungraciously, 'In 1844 the church was restored by Canon Simpson, Miss Losh, and others, and has been since rather unfortunately enlarged.'[15]

Sarah Losh (1785–1853) was an extraordinary woman: intelligent, practical, artistic, attractive (judging by her portraits), full of intellectual curiosity. She was partly self-taught, partly taught by her uncle and tutors; proficient in Latin and Greek and modern languages, the Classics, algebra and geometry and knowledgeable about geology and fossils; she learnt to model in clay, carve wood and sculpt stone; she was an 'architect-designer', an estate-manager, a philanthropist – and apparently widely liked and admired. Her family were friends with the Wordsworths, Coleridge, Southey, and engineers such as George Stephenson; they owned an alkali factory on Tyneside and were involved with the building of the Carlisle–Newcastle railway. Her life is celebrated in Jenny Uglow's authoritative biography, *The Pinecone*[16] – but Sarah Losh is a woman who should be celebrated even more widely as an example of what a determined woman can do. It helped, too, that her family – who lived at Wreay (pronounced *ree*-uh) just to the south-east of Carlisle – were well-off, and perhaps also that she remained unmarried. Sarah had relatives who lived at Burgh-by-Sands on the Solway, and earlier generations of Loshes or Arloshes had been 'grangers', farm managers, on farms belonging to Holme Cultram Abbey. Sarah, her sister Katherine, and various of their Newcastle aunts and cousins occasionally visited the West Cumberland coast, staying at the coastal spa of Allonby. Sarah was apparently shocked that the locals were robbing the decayed Newton Arlosh church of stone for their own houses and, in her fifties and already proficient in architectural design and construction, she decided to rebuild it. She used local materials wherever possible in all her buildings, and employed heavy flags of Lazonby New Red Sandstone, quarried in the Upper Eden area near Penrith, in the renovation. But when you walk up the path towards the church it is the hunched, observant eagle on the roof that strikes you. This eagle – unlike the eagle at Holme Cultram, which is poised to fly off towards the sunset – stares out to the east, wings folded. Inside the church are sculpted rams' heads with curling horns, and a lectern carved like a palm tree, probably by William Hindson; the font, which was carved from bog oak, has been stolen. The pale sandstone sheep are stylised, not representative of the local breeds: it is thought that Sarah carved them, and the eagle, herself. The church is comfortingly sturdy, sitting close as it does to the creeks and tiers of Newton saltmarsh. Although it shows signs of Sarah Losh's ideas and work, it is very modest in comparison with her exuberant and astonishing Church of St Mary's at Wreay. A great deal has been written about this treasure, by architectural critics such as Nikolaus Pevsner – 'The remarkable creation of Sara Losh . . .'

She created Wreay church in 1842 in memory of her beloved sister Katherine. Wherever you look, inside or out, there is carved wood, sculpted stone and richly glowing glass. It is a geologist's and zoologist's delight: many of the animals and plants, whether depicting fossils, living species or mythical beasts, have hidden meaning and significance. Lazonby sandstone, bog oak, oak from the Losh family's woodlands – and her fascination with the fossils found in the shale bands in the Cumberland coalmines – all contributed to her extraordinary work. She carved alabaster to make the lotus flower on the font, and cut thin sheets of that transparent stone to make narrow window 'glass' with silhouetted fossil ferns. The design of the Church of St John the Baptist at Newton Arlosh makes perfect sense in the context of its restorer's skills and her connections with the Solway and Holme Cultram Abbey.

*

I visit Newton Arlosh again to go out on the saltmarsh to learn about stints. The marshes of Burgh, Skinburness and Calvo, and Newton Arlosh and Saltcoats are grazed by cattle and, traditionally, by hoggs of the famous fell sheep breed, the Herdwicks, that were brought down to fatten on the saltmarsh grass. Now the breeds are more diverse and the grazing rules are strict, for the marshes are divided into stints. The stints are privately owned but unfenced, and so the marshes are 'shared grazing' – the letting of which is auctioned annually. It sounds quite simple, but as Eileen Bell, then Secretary of the Newton Arlosh and Saltcoats Marsh Committee, explained to me, it isn't.

In October I drive along the edge of the Firth under an orange-yellow sky louring over the flatlands, its light dimmed by Sahara dust, the red ball of the sun showing fleetingly through the hurrying clouds. But the forecast rain hasn't appeared, and when I reached Eileen's house she at once suggests we get booted up and go out onto the Marsh. We head west in her Land Rover through the village, and down a lonning[17] past the barns and red sandstone walls of Orchard House where she was brought up, and past another farm by a large pond where mallard and a moorhen scull undisturbed. Eileen's family have lived in this area for three generations; she tells me her paternal grandfather 'worked tire-lessly for the Marsh'.

We park by a gate that leads onto the Marsh. The entrance has clearly been a favourite gathering-ground for cattle, as it is poached by their hooves to a slurry of ankle-deep mud. Clinging onto the fence we teeter round the edge and onto the short turf. The Firth is only visible in the distance as a thin silver line. Newton Arlosh Marsh borders the River Waver and the River Wampool where they open into Moricambe Bay, and its extent is much greater than I had previously imagined. There is no sign here of the distant creeks that carve deep muddy fractals through the Marsh's outer edges. Although well-grazed, the landscape is not monochrome, but a palette of greens and ochres. The areas that are covered by the big spring tides are paler green than where we are standing, a different sward with different, salt-tolerant grass. By the gorse-covered raised ground at the top of the Marsh there are large tree trunks that have been carried in by the overtopping tides. Further along, we come to a new fence that heads out towards the Firth. Unlike Burgh Marsh to the north-east, where the Hammer of the Scots' memorial stands, the whole of this marsh is fenced, even at the water's edge, to lessen the probability of the cattle becoming mired. That's important. Eileen says she can remember her father having to dig cattle out of the creeks with ropes and spades. 'Before it was fenced, cattle would get out onto the sand at the [River] Wampool. I've crossed the Wampool to Anthorn! The cattle were out on the sands, one night after supper, and we had to go down and get them. We herded them to the other side, and then we had to get permission to put them in a farmer's field that side. And then we had to go back and get them in the morning.'

Our next stop is on the edge of Middle Marsh, nearer to the village. Young steers come galloping across a field to snort and snuffle by the gate, perhaps hoping that our Land Rover, with sacks in the back, means food. We squelch down a track between two hedges, one trimmed to almost suburban neatness, the other rich with red, wizened haws and deep purple sloes as large as damsons. A heron extends his neck and lumbers into the air. Again, the Firth seems far distant, and scattered cattle are indistinct specks. There is no sound other than the faint hissing of wind in the grass. Eileen tells me how she has always loved being out on the Marsh, by herself and with her children; her face lights up as she points out several hundred starlings, previously hidden, which have lifted off the grass and now perform a small murmuration before settling again. And now, too, the yellow-orange sky suddenly clears, and the sun is warm on our faces: colours brighten and the Marsh comes alive.

Picking our way back to the lonning we pass large circular water troughs by the gate, and though the grass is now wet enough that the cattle don't need to drink, nevertheless the ground has been churned to sloppy mud, through which a quad bike and a very muddy border collie are splashing towards us. Eileen introduces Steven, the Marsh Herd – he tells her that he's decided to move the cattle off a couple of days early because big tides and high winds are due (ex-hurricane Ophelia is blasting in towards us from the west). Later, from Eileen's house, where we are having tea with scones and bramble jelly, we can see the distant cattle starting to move westward in a line. I can just make out the quad bike chivvying them, darting round them like a cattle-dog. One beast with divided loyalties breaks away and heads back towards a group that hasn't yet started to move; Steven circles and sets them moving, galloping after the others. 'He'll be moving them to the pens. There's a big ditch down the middle of the marsh, but there's a bridge, and they'll know where to go, they've been out there since May.'

The owners of the Marsh's stints meet once a year, and decide who will be on the Committee. When I meet her, Eileen is secretary, her husband, Willie, is chair, and their son and two others make up the other members. Eileen says, 'I first got involved when I was about nine. I was given the balance sheet to type out. We did it with carbon paper in those days. Mum would do a lot of the writing' – she shows me a book filled with neatly written notes – 'and my sister was secretary in 1974.'

The area of the Newton Arlosh saltmarshes is 440 hectares (about 1,100 acres). So, what size is a stint? I now discover that a stint isn't a set area. The measure depends on the grazing offered. 'One stint can be let as a "stint-and-a-half" because of the abundance of grass.' Eileen laughs and shakes her head at my expression. 'So if I have three stints, I let them as four-and-a-half. But it's a temporary measure, it depends on the grass.'

In other words, as Winchester and Straughton explain in their paper on stints and sustainability, stinting refers to the 'carrying capacity . . . a notion of the total number of animals that should be allowed to graze there'.[18] The carrying capacity clearly varies depending on circumstances, such as the abundance of the grass, as Eileen explained. To an outsider the calculations seem complicated, but to the graziers it makes perfect sense. For example, on Newton Arlosh, two cattle are permitted per stint. They must be heifers or bullocks – cows in calf are not permitted, not only because of the danger of contagious abortion, but also because it would be difficult to deal with an awkward calving out on the Marsh. Eileen checks her book and tells me that in 2017 the available stints have been let to twelve farmers, with from six to sixty-three head of cattle. No sheep are permitted on the Newton Arlosh and Saltcoats Marsh. The Marsh Committee of Skinburness and Calvo Marsh, in contrast, allocates either one head of cattle, or two ewes and four lambs, or four geld sheep that are not in-lamb per stint (and after 1 August, this changes to four sheep or four lambs per stint). At Burgh Marsh, the grazing is not as good, so the allocation is only one head of cattle to a stint. The stints are auctioned one evening in late March each year, and the 'stinting day', when animals may be brought to graze, is usually in the first week of May. Animals must be gathered in from across the marshes and removed, with the help of the Marsh Herd, in mid-October. But since these saltmarsh graziers are dealing also with the Solway Firth, those start and finish dates are dependent also on the tides and weather. Ultimately, the Firth is in charge.

'I'd rather not disclose.' Eileen smiles to take the edge off her refusal. I'd asked how many stints there were on Newton Arlosh Marsh. When I probe a little more, it seems that the availability of stints affects the price, so the numbers are not disclosed even at the auction. At one time the price rapidly increased because it seemed that word had got around that cattle did well there because the grass was rich – no expensive fertiliser was needed and the salty grass reduced the likelihood of infection by parasitic worms. I asked an acquaintance of mine who works for the auctioneers if she could explain the secrecy about the number of stints for auction. She laughed, and told me, 'It's a strange rule! We're never allowed to tell anyone on the night how many stints we're auctioning. We can give little hints, like "not many left", or "we're down to the last few" . . . On Newton Arlosh, we auction them in lots of ten; it's twenties on Skinburness and Calvo.' Moreover, the number of stints available to auction changes every year. 'It depends, some owners graze their own, but others decide to let them. Some of the stints have been passed down for generations. A lot of the owners are, shall we say, senior citizens, but there are some younger ones too. The number is set by the separate Marsh committees. We don't know until the night.'

The Bells breed pedigree Friesian cattle, and Eileen and Willie were founder members of the British Friesian Breeders' Club. Earlier, we had talked about the impact of the 2001 epidemic of foot-and-mouth disease. Even now, sixteen years later, Eileen clearly found it a very difficult topic. 'We'd built up the herd over forty years,' she said. 'They all had to be killed, two to three hundred animals. I remember Willie bringing the bull – he was a big bull and no one else wanted to do it – out of his pen to be shot. I think that's when Willie lost heart . . . But we had some semen stored. And then other members [of the Friesian Breeders' Club] started getting in touch when they heard what had happened. We were offered brilliant stock – it brought a lump to your throat to think they'd give you such stock.' We were both quiet for a while as we remembered those terrible months of killing throughout the county. My husband and I moved to our smallholding within sight of the Solway just two weeks before the epidemic struck and for the next eleven months we had painted wooden sheep in our field; we were fortunate that we hadn't already bought and lost treasured stock. But I will never forget the image of a digger on an overbridge, its bucket silhouetted against the sky as it poured a stiff-legged tangle of sheep carcases into a waiting wagon. Our own village was 'cleared' on Good Friday.

Watchtree Nature Reserve is on the site of one of the Solway's several wartime airfields, Great Orton, which, abandoned after World War II, became an unofficial 'community resource' used by microlite flyers, clay-pigeon shooters and youngsters learning to drive. But in February 2001 it was commandeered by the army as a mass grave, an enormous engineering and logistical feat that saw twenty-six burial 'cells' dug – and filled. The name, Great Orton, became synonymous with 'the killing field', the great trauma of the county, which many find hard to talk about even today. There is a plaque at the gate, mounted on a piece of Criffel granite, a glacial erratic that was dug up nearby:

A Symbol
To the birth of
Watchtree Nature Reserve
Dedicated on this day the 7th May 2003 on
the second anniversary of the final burial.
A Memorial
To 448,508 sheep, 12,085 cattle, 5,719 pigs
buried here during the
Foot and Mouth outbreak of 2001

Typing those words makes me want to weep even now, and I try to push the stories and images from my mind. The Bells lost their cattle; at Rockcliffe Marsh Giles lost most of his sheep as well as cattle, and parts of the Marsh grew a new crop, hay.

But, ungrazed, the saltmarshes flowered. In her internal report for that year, the Wildlife Trust's summer warden at Rockcliffe wrote that

Every large creek was edged with the longest vegetation, typically tall grasses and Meadow Buttercup (*Ranunculus acris*), in addition to Common Birds Foot Trefoil (*Lotus corniculatus*). This gave the impression of yellow rivers running through the Marsh, as they followed the creeks . . . Further down, a lilac swathe of Yorkshire Fog (*Holcus lanatus*) covered a strip from the top of Yellow Creek to the Eden . . . [at the Fleam] the creeks were edged with daisies (*Bellis perennis*), Thrift (*Armeria maritima*), grass which was calf-length in height, tall buttercups and patches of mid-thigh length Spear Thistles (*Cirsium vulgare*) . . . On leaving the Marsh, heading back across New Bridge the final impression was of First Field, apparently a mono-crop of the white, daisy-flowered Scented Mayweed (*Chamomilla recutita*), divided only by the path in the middle.[19]

And Brian Irving, former manager of the Solway Coast AONB, told me, 'It was the first time that I saw the marshes really flower, you know. The first time in their history of grazing – an increased level of genetic diversity going on because the plants were allowed to flower and set seed, rather than propagate by roots and suckers. Stunning, the quality and texture of the marshes. A once-in-a-lifetime sight – and it was there for all the wrong reasons.' The Solway's saltmarshes were transformed during the summer of 2001. The sky above Rockcliffe Marsh was loud with the singing of larks.

*

Were the marshes transformed during wartime too? On that bitterly cold, windy March day when Adam, Andy and I went to look for mudshrimps, *Corophium*, at the edge of the Nith, we also found some inanimate artefacts of former times – five empty shell casings, tangled amongst the roots of a fallen 'merse-berg' and half revealed by the tides. After Andy had washed away their thick muddy coating in a pool, we could see that the cases were each about fifteen centimetres long, still shining coppery-brown beneath the patches of bright blue verdigris. Stuart James, the archivist for the Dumfries & Galloway Aviation Museum, thought they could be cannon-shell cartridges, which would have been fired from aeroplanes at targets out on Blackshaw Bank just off Caerlaverock, 'most likely by Hawker Typhoons from any of the RAF's Operational Training Units such as Great Orton and Crosby-on-Eden'. Blackshaw Bank was a practice firing range, he told me, for more or less the whole of World War II, 'especially during intensive training leading up to D-Day'. I wonder what effect this noise and disturbance had on the migrant geese and waders – if they moved away from the Solway's shores during those years. If so, then the flowers of the merse, ungrazed by geese, perhaps bloomed more wildly in the summer; and the mudflats' invertebrate populations, less preyed-upon, perhaps exploded in numbers too.

*

At Campfield Marsh, Guillaume (Will) Goodwin is finishing his field-work as he nears the end of his PhD. From Normandy, but with only the slightest of French accents, he is based at the School of Geosciences at the University of Edinburgh, and has been studying the changes in the topography and sedimentation patterns of the saltmarsh each side of the stub of the Solway viaduct's embankment. This weekend, at the end of March 2019, he has brought two colleagues, Louis and Marie, to help him, and we are all muffled in layers of warm and weather-proof gear against the bitter north-westerly wind that has been blowing for a fortnight.

Today Will is recovering the sedimentation plates that had been anchored two years previously in the top of the Marsh on both sides of the embankment. He laughs, slightly ruefully: 'Actually it's very emotional – Louis is reclaiming his soul!' Apparently Louis, who is also a postgraduate, had helped put the plates in place two years previously, on a weekend when sleet was blasting across the Marsh. Today might be cold, but Louis assures me it is almost mild compared with that weekend. The flat metal plates had been fixed in place with iron pegs; at first they were marked with sticks topped by fluorescent tennis balls, but these didn't survive; the next attempt, stakes with high-vis wrappers, also disappeared, due to human or cattle intervention, or the storms. Fieldwork is rarely straightforward, especially when it involves interacting with the sea. But now Will sets up the yellow tripod that supports the GPS equipment and notes the accurate coordinates that will allow him to place the plate in the computerised model later. Louis walks across the slippery surface to find it. Marie, a French intern, does the cold, damp work of scraping the accumulated sediment and vegetation into a labelled bag. The sample will be weighed and the particle size measured back in the warmth of the lab.

Will is interested in changes in the height and shape of the Marsh, especially at its lower end, rather than the chemical content – such as stored carbon – of the underlying sediment. I wonder how he maps this, and Louis laughingly tells of struggling with the department's laser scanner across the Marsh – all twenty kilograms and £100,000-worth of equipment. The scanner moves through 360 degrees and measures the topography, but because it cannot always 'see' the creeks it has to be moved to different positions. In theory, and depending on the vagaries of software, the scans are then stitched together by a computer program that constructs 3D images for comparison over the fourth dimension, time. Will tells me that the Marsh is growing vertically on both sides of the viaduct embankment, although the upstream sites east of the embankment have slightly receded. It seems that as long as the vegetation stays in place, the Marsh is fairly impervious to changes in the tidal mudflat. The sediment deposition generates a gentle slope rather than a wall, which softens the action of waves. Today, he finds that all of the test plates have been buried by sediment to varying depths, and one on the downstream, western, side of the embankment by as much as a surprising eleven centimetres.

After they have retrieved all the plates, we head across the mud and climb up the sloping sandstone blocks of the embankment; I have both hands full carrying plastic boxes of research kit, and climbing in wellies up the lower stones that are coated with green sea lettuce, *Ulva*, is an interesting challenge. Partly hidden by the gorse and scrub that form a scruffy fringe along the embankment's smooth brow is a rusting mussel sorter, dating from the time of a mussel farm which proved uneconomic nearly twenty years ago. This embankment has been a witness to doomed ventures (see Chapters 2 and 6, for the Solway Railway viaduct). Next to the mussel sorter is the equally incongruous yellow tripod of the GPS base station, where we dump most of the equipment and slither down the dressed sandstone blocks – the work of Victorian stonemasons in the 1860s – to eat our lunch in the lee.

*

Dig down into a saltmarsh and within a few centimetres the sediment is darker and compacted. As with peatbogs, this deeper layer is anoxic, depleted in oxygen, and where there is little oxygen, few species of animal or bacteria are able to live – or to eat or digest dead and buried plants and their roots. Plants and single-celled diatoms take in carbon dioxide from the air and convert it to other molecules and energy; normally when they die and have decomposed, the carbon is oxidised again and released. But as a saltmarsh or peatbog grows upwards, the dead plants, diatoms and other organic materials are compressed and preserved in the anoxic conditions. A saltmarsh can act as a carbon sink, so its stored carbon is no longer available to be converted to carbon dioxide and released into the atmosphere.

The rôle of estuarine saltmarshes in mitigating climate breakdown is now well recognised, and there are many research groups around the UK and abroad that are devising experiments out on the marshes and in the lab to study this. Research has shown that stable Welsh salt-marshes store four times as much carbon as marshes that erode and accrete in an unstable fashion. Other work, at St Andrews Sediment Ecology Research Group (the 'Mud Lab'), shows that a saltmarsh stores no more carbon than the adjacent mudflat.[20] In other words, no sweeping generalisations can be made.

On a bright, windy day I stood near the edge of Calvo Marsh as a big tide raced in, and watched as the waves smashed against the fretted edges, roiling into the creeks. The power of those lumps of water fractured as they were transformed into spray. As the tide rose up through the creeks and spread over the Marsh, the colours changed too – from brown and frothy white to a sheet of blue. For what is obvious is that the bumps and hollows of marshes also play a large part in dissipating the energy of waves and in flattening storm surges; they play a similar rôle to that of the floodplains around our rivers. As Norman Nicholson writes in *Cumberland and Westmorland*, 'They are neither land proper, nor sea proper': they mediate between the two.

Finally saltmarshes are starting to attract public attention, and the experience of visiting a saltmarsh and walking across it, negotiating it, can be a revelation. I have run creative writing days for mixtures of beginners and practising writers out on the marsh at Campfield (we also, fortunately, have the option of retreating indoors into the RSPB Solway Wetlands Centre – the weather can sometimes make the visits 'very atmospheric'). We talk about how the Marsh is formed and its importance, but we also wander around and look at the plants and animals, the hummocks and the creeks (the limestone pavements that encircle the Lake District Fells have much better words – clints and grikes – for that intersected flatness). Then everyone is left alone for a while, to sit or walk, to look, to write. This is the part of the writing day that makes me happiest: when I can watch people becoming absorbed by, and briefly integrated into, this strange not-land, not-sea. The reactions differ not only from person to person, as you would expect, but also from year to year. Initially the response is of surprise, and sometimes even discomfort: 'There's a strength to the landscape, a sense of danger'; 'The Marsh makes you walk where it *wants* you to walk'. But others instantly feel at home. Gradually the writers start to enjoy the differences in scale, from the wide landscape to the minutiae.

*

It turned out to be the hottest day of the year when the Biodiversity Officer for Dumfries & Galloway, Peter Norman, led me on an 'expedition' across Kirkconnell Merse. Some while previously, geomorphologist David Smith had told me about a trackway of birch branches which he had found when studying changes in sea level along the River Nith; the traces of the trackway across Kirkconnell Merse led to the gravelly crossing of the Nith near Kelton, and he speculated that the trackway might date back to the Bronze Age. Unfortunately, its coordinates hadn't been noted, nor was it mentioned in any records – but Peter got in touch to tell me instead about the Nith 'training wall': a wall that had been constructed to train the unruly river to know its place. Our rendezvous was in the pretty village of New Abbey at Sweetheart Abbey, now a ruin of red sandstone partially wrapped in scaffolding, but with the complex stone arches of its windows – trifoliate, rounded, pointed, and a huge rose window – still intact. A Cistercian abbey, like Holme Cultram, it was founded in 1273 by the Lady Devorgilla of Galloway after the death of her husband, John Balliol, the father of a contender for the Scottish throne, in 1269. A faded information board explains that she had her husband's heart embalmed and placed in a casket, as her 'sweet, silent companion'. When she too died, she was buried in the abbey with the casket 'clasp'd to her bosom' and the monks named the abbey *Dulce Cor*, 'sweetheart'.

We park near Kirkconnell House and set off down a well-made track that runs parallel to the Merse. It is a tunnel of trees, providing welcome shade for us and for dozens of fritillaries, their orange-brown wings flickering in and out of the shafts of sunlight. The largest tabanid fly I have ever seen, like an overgrown cleg, is resting on a stone and I give it a wide berth to avoid disturbing such a huge piercing, blood-sucking machine. The Merse is a gently swaying field of green, the seed heads of its grasses straw-pale, and to reach it we must climb down through a bramble thicket and under barbed wire, then into and across a shallow trickle of water in a creek. Peter has been here before and knows that we will have to make diversions around several large creeks. This is old merse, well-established and dense with thigh-high grasses that spray clouds of pollen, and speckled with the flat white heads of yarrow. It is merse that has grown upwards, so that every creek is deep and slippery-sided, too big to jump across; it 'makes you walk where it wants you to walk'. Two little egrets and I are hidden from each other; they see or hear me first, and unpanicked, unhurried, but very purposeful, they lift into the warm air, their wings dazzling in the sunshine. Necks stretched out and long black legs trailing, they fly towards the river. They have left nothing but footprints, slim, elegant and long-toed.

Progress is slow as we look for easier routes, climb down into creek bottoms sticky with mud, climb out again. The temperature out on the Merse is a scarcely believable 30° Celsius: it's hot, sweaty work, and there are clegs. Gradually we reach the band of shorter saltmarsh vegetation – new plants for me, fleshy orache and the spikes of plantains, as well as scurvy-grass, several asters already showing pale blue flowers, arrowgrass and even large patches of samphire. There are bare areas, too, where the mud has cracked into a crazy-paving of polygonal shapes with raised edges, or where the footsteps of a fox and waders have been baked like trace-fossils. Sheep-like, we follow a yellowish line of a sheep-trod and eventually we come to the merse-edge, and although there are still the branching runnels of creeks to negotiate, the gravelly mud makes easier walking as we head south towards the river mouth. Now at last there is moving air and space to breathe; six bright white egrets fly ahead of us, and a pair of eider ducks are *oooh*-ing softly out on the water; swallows swoop and a cormorant beats rapidly down-river. On the opposite bank is the merse at Fishermen's Bush where I dug for the mudshrimps, *Corophium*, and found World War II shell casings, in weather that could not have been more different.

Peter and I step carefully down the sloping mudbank to look back at the Kirkconnell Merse and from here the three tiers are obvious, with different colours, different textures, the face of the nearest and lowest striated with sequential deposits by the tides. And beneath my feet is the densest colony of *Corophium* that I have ever seen – the tiny cones at their burrow mouths are so close-packed that there is scarcely room for more. Now I can believe those estimates I read of 10,000 or more per square metre. I wish I had a spade to look at their colony more closely. But we have to keep on moving south before the tide turns, to reach the training wall. On my left there is a raised line of dark rocks, jumbled together along the river's edge. Between this 'wall' and the merse to my right is a wide muddy slope, smooth and glistening in the sun, where the mudshrimps live and where waders and gulls have left their busy footprints and probe-marks. The barrier of stones becomes higher and more defined; although previously the rocks had been coated with mud and thin, flat fronds of green *Enteromorpha* algae, these are black and sharp-edged, a type of local greywacke rather than the granite that I had expected. And then we finally reach the iron rails of a narrow-gauge track that runs along the top of the wall. In places the rails are still attached to wooden sleepers, elsewhere they have been bent and jut upwards like spars, or hang across eroded hollows like a miniature version of the ice-battered Solway viaduct. A hollow metal post sticks up on the seaward side, its rusty outer surface coated with barnacles. Yellow-brown fronds of *Fucus serratus* and channel wrack, *Pelvetia*, dangle over the track, and even up here there are the burrows of *Corophium*, the mudshrimps taking advantage of mud trapped between the rails. Now, even at low tide, the Nith dominates; it ripples down towards us, a ribbon of blue and brown, edged on the far side by sloping mud that reaches back towards a narrow green strip of merse, and on this side the rails draw a line between the water and the mud and merse. We are all confined within this quiet river-sculpture by the borders of distant trees. It is a wonderful and surprising place.

Until the early twentieth century the rivers Nith and the Annan were important not only for their ports for incoming and outgoing goods and people, but also for shipbuilding: the Nith was the gateway to the port of Dumfries and Kingholme Quay and 'out-ports' for smaller ships like Glencaple and Kelton, while the River Annan led to Waterfoot, Welldale and Annan. But, like the other rivers with shallow, estuarine mouths that open into the Solway, the rivers have always been prone to silting up due to soil carried down by the rivers themselves from inland, and sediment in the incoming tides. The channel of the Nith also meanders this way and that across its width, and by the early 1800s the passage of ships was becoming increasingly difficult – the Nith needed to be restrained, confined to a narrower course that would increase the energy of the flow, and flush away the sediment. As Walter Newall wrote in 1847, 'If the line of the channel were fixed as shown by the plan, it would gradually be deepened by the influence of the currents, and if aided by artificial means, its progress in deepening would be very rapid.'[21]

Peter Norman has already carried out some research into the Nith Navigation Trust, which was responsible for commissioning the building of the training wall, and he generously shared some of the information with me; I have picked out only a few themes here since he will be writing in more detail elsewhere. A survey of the Nith by the engineer James Hollinsworth in 1811 led to suggestions for improvements, and over the next few years several surveys of the river were carried out by Walter Newall, who was not trained as an engineer or architect but had come into this work from designing and building furniture. He produced plans for the Nith Navigation Trust for restricting the line of the river, and the engineering practice of D & T Stevenson, of the famous 'Lighthouse Stevensons', was instructed to produce plans for a wall that would confine the meandering river to a straight channel past Glencaple. Peter thinks the work was carried out sometime after 1860; it is possible that the stone was brought in and dumped from barges, and – when the tide was too low for them to approach – transported in wagons along the rails on top of the growing wall. The wall that we walked along, and which the mudshrimps and algae had colonised, was more than 150 years old. Now it lies along the western margin of the river, abutting the Merse – but when it was built, the estuary was much wider, and the channel of the Nith was over to the west of the training wall; in other words, the wall must have been built out in the middle of the turbulent estuary, despite the tides and weather. This really points up what a dangerous job this must have been – and puts the later Solway viaduct construction in context: the Victorians unafraid to tame the Firth, yet again!

We had walked past Kirkconnell House with its parkland of ancient gnarly limbed oaks, and from there we had struggled at least a half-mile across the Merse as the crow flies; but the house had once been close enough to the river to have its own jetty. The 1850 Ordnance Survey map shows the jetty, no merse, and of course no training wall, but fifty years later the updated map marks the line of a wall and its posts, the growth of the merse to the west of the wall – and no jetty. The Nith had been pushed to the east. The Kirkconnell Merse has accreted and grown upwards, protected by a wall for about 180 years; it would make an interesting research project for a student of saltmarshes.

The putative Bronze Age trackway across the Merse that David Smith had mentioned was still elusive, so two months later I followed up another of his suggestions, to choose a low spring tide and walk up the merse on the eastern side of the river to the 'ford' at Kelton. Now I walked to the accompaniment of a soundtrack of rain pattering, sometimes thudding, on my hood; my splashing footsteps; the occasional *phweet* of lapwings and the piping of redshanks. A curlew was bathing in the shallows, dipping and raising its head, and fluttering its wings. I learnt recently that the curlew's long beak bends and flexes as it pulls worms from the mud. Twice I disturbed the flock of fifty or so lapwings as they rested or probed the mud for snails and mudshrimps with their short beaks, twice I apologised as they rose up and circled. A greater black-backed gull was lurking thuggishly near a fishing heron but flew off as I approached. Out on the water a cormorant was struggling, head raised, to swallow a fish; it reminded me of a dead cormorant we had found on a Norfolk beach, which with its fish prey must have died a slow, ironic death – jaw and beak of prey and predator interlocked. Rain showers swept up the Firth, intermittently blotting out the hills, but the Nith was flowing smoothly beneath muddy banks that were intercut with branching runnels. Was there a ford? I had asked someone in the café at Glencaple but she had never heard of one, and there was no one else here to ask, not even a dogwalker. The distant person in a red coat standing watching the river turned out to be a post and lifebelt. But at a bend the water's surface was agitated by small argumentative waves, and here it was clearly flowing over stonier ground. The riverbed was wide and flat, and it was possible to imagine animals and carts being driven down the sloping bank behind me. Opposite, though, the edge of the Merse was high and although I scanned it with my binoculars, no trackway was revealed; unsurprising, since there would have been such great changes in the topography over that period. However, there was a short double row of posts projecting from the mud and disappearing back into the middle of the bank – perhaps the remains of another jetty, or even the posts of former stake-nets; their story probably forgotten, yet still intriguing.

A letter by Walter Newall in the Tidal Commissioners' Report of 1847 notes that 'the port of Dumfries extends from Dumfries to Southerness Point' and of the eight ports which that includes, 'six are within the river. Glencaple, a tidal pier; Kelton an open scar; Kingholm, a small tidal basin and wharf wall; Laghall, a small tidal pier; Castledykes, a small tidal wharf wall; Dumfries, a small tidal wharf wall.' Dumfries imported so much tobacco from America in the eighteenth century that it was known as the Scottish Liverpool, according to the Ports and Harbours of the UK website: 'Coal from Cumberland formed the largest percentage of the trade, and continued into the twentieth century. Lime was another common import, but Brandy, wine, dried fruits and luxury textiles, came from France and Spain. These together with tobacco were the commodities that gave the smugglers their trade. The customs accounts of the time relate what a hopeless task it was to control it.'[22] Indeed, the Solway was infamous for its smugglers. Luxury goods from America and Europe were imported by dealers in the Isle of Man, then repackaged and sold on to the smugglers; the rivers and coves of the Scottish Solway shores were especially suitable for landing the contraband. The *View on the Solway, Mouth of the Nith, Criffel in the distance*, painted in about 1850 by Clarkson Stanfield, shows a line of ponies waiting by a beached two-masted boat, and in the foreground a black-and-white dog guarding a pile of barrels that are part hidden by a sail. Robert Burns, as excise man, was probably kept busy; he reputedly composed the verses of the song, 'The Deil's Awa Wi' The Exciseman' (1792) as he kept watch on a group of stranded smugglers near the River Sark while a fellow exciseman went to get help to board their lugger:

> The deil's awa, the deil's awa,
> The deil's awa wi' the Exciseman,
> He's danc'd awa, he's danc'd awa,
> He's danc'd awa wi' the Exciseman.

Glencaple still has a pier (and popular café), but it's no wonder that there remains little sign that Kelton, 'an open scar' (referring of course to the rocky scaur rather than the hamlet's appearance), was once an out-port with a small shipyard for building and repairing boats. Along the Nith, mud and saltmarsh define the landscape. I drive back north to Dumfries, stopping off at Kingholme Quay on the way to see the red sandstone tidal basin, where several yachts and a catamaran slouch on the mud waiting for the tide to return. A beam trawler, the *Petronella*, rests lopsidedly against the wharf wall a little further downstream, its propellor completely exposed; this large boat made its way upstream in 2005 and has never left.

Walter Newall's engineering work for Nith Navigation allowed the town of Dumfries to prosper as a port. He was born in 1780 and died eighty-three years later at nearby New Abbey, but he spent his working life in Dumfries. There he is better known as the architect of many of the town's handsome buildings and for his conversion of the Maxwelltown Windmill into an observatory, with telescope and *camera obscura,* for the Dumfries and Maxwelltown Astronomical Society. On the first floor is the tall brass eight-inch telescope that the Astronomical Society commissioned – along with the fittings for the *camera obscura* – from Thomas Morton of Kilmarnock, who had taught himself to make scientific instruments. The telescope's design had to be approved by the 'local boy' and Arctic explorer Admiral Sir John Ross, a man one might presume to have some practical knowledge of the efficacy of spyglasses, before the Society accepted it. It is a beautiful piece of equipment about five feet high, its polished tube pivoting on tapered brass cylinders. The brass-bound, mahogany three-legged table to which it is fixed can be wheeled to whichever of the long windows the viewer requires. A piece in the *Dumfries Courier* of August 1863, the month the observatory was opened to the public, tells how 'In a tolerably wide range of country, every gentleman's house may be examined outside.' Not just outside, either: 'On Friday, when the telescope was pointed to the lower part of the town, a lady could be seen sitting at her window reading a letter and when this was mentioned, Mr Morton remarked that by putting in the most powerful glass, he could enable the beholder to read the letter too.' Moreover, viewers could now study the river banks and merses, and 'vessels described afar off, standing for the Nith, the Annan, &c, and the deck passengers on the Nithsdale counted, if not distinctly recognised, when she sails for or returns from the port of Liverpool'.

*

At Campfield Marsh on the Cumbrian side, on that cold March day when I meet the research students, the Solway is as brown as milky tea, the wind whipping the spray off the white-topped waves; at my back the sounds are bass and treble, a low roar of waves breaking against the embankment, overlaid with the whining of wind in the few remaining pillars of Victorian ironwork. Waves sweep around the end in a great arc and roll in on the eastern side, filling the deep creek in the mudflat beneath us. In five minutes a stone-scattered mudbank just offshore becomes an island; in fifteen minutes it has vanished. As the mudflats separating us from the saltmarsh are quickly overrun by this 'mucky tide', I try to imagine how the mudshrimps are responding. Even on the mudflats the water seems too disturbed for them to leave their burrows. Ten or eleven thousand years ago the cold post-glacial rivers were flushing gritty rock débris down to the Irish Sea, and the comforting cushion of mud had not yet developed.

I pick my way down to the water's edge, reaching out to scoop a handful of water that is thick with fine brown particles. And I see that the mud and the marsh are inseparable; they are the integrated margins of the estuary. One gives rise to the other; damage to one leads to damage to the other. It is as though they are holding hands to keep their foothold on the edge of the land.

5

Peat

The Solway Firth is unusual amongst firths in that it is so closely linked with peatlands and on such a scale. Only a little way inland from the salt-marshes are the Solway's precious 'lowland raised mires', otherwise known as peatbogs or, more poetically, Mosses or Flows. Because their age is so great and their development so slow and fragile, and because the plants and animals that live on the Solway's Mosses are so intimately adapted to their home, most of them have statutory protection and are managed by various conservation organisations. But they are also wild and beautiful, appealing to all the senses. They remind us, too, of our significance to their survival.

'The Mosses vary season to season,' Frank Mawby told me. At that time he was the Northern Reserves Manager for English Nature (now Natural England). 'In winter, they're brown, not surprisingly, but they're still the most colourful habitat you can walk on. When the *Sphagnum* mosses are all wet, the different species have different colours, greens and oranges that are almost fluorescent, they glow. Big hummocky ones that are dark red . . . All these under your feet, under the layer of dead cotton grass. A skin of mosses. It's very quiet at times, but at other times it's pretty hectic – snipe, jacksnipe which jump up under your feet, pipits and skylarks. There are the bloody midges, too!'

Frank's office, a strangely ramshackle building with a sheet-metal door, is at the end of a potholed track that is also shared with unnervingly large lorries that swing out of a small industrial estate on the old

Looking north across the Solway to Criffel and Scotland: Moricambe Bay (foreground), Calvo and Border saltmarshes and Grune Point (© James Smith)

Bowness Common, looking south to the Lake District fells. Campfield saltmarsh fringes the coast, and fields edge the seaward side of the Common; the part-flooded line of the old railway track crosses 'the Moss' (© James Smith)

Sun and snow showers

Robin Rigg windfarm seen from the Galloway coast, with the Lake District fells
in the distance

A spadeful of burrows of the mudshrimp, *Corophium*, surrounded by yellowish oxygenated mud

A heron by Hanging Stone, amongst the reefs of the honeycomb worm, *Sabellaria*: Allonby Bay, Cumbria

Dog-whelks, *Nucella*, laying eggs

Banded snails, *Cepaea*, on the dunes

The tubes of the honeycomb worm, *Sabellaria*

Cages and barnacle-encrusted posts of the former oyster-farm, Dubmill Point, Cumbria

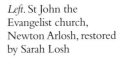

Left. St John the Evangelist church, Newton Arlosh, restored by Sarah Losh

Middle left. Holme Cultram Abbey, Cumbria (Ashley Cooper pics/Alamy Stock Photo)

Below left. The Reading Room at Allonby, Cumbria

Below. Ruins of Sweetheart Abbey, New Abbey, Dumfries & Galloway

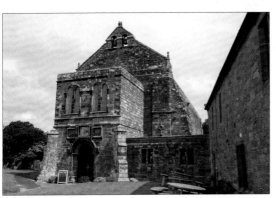

The cliffs at Rascarrel Bay, Galloway

The submerged forest and peat-banks near Beckfoot, Cumbria

Above. Raising the oak roof timbers on the clay dabbin house at RSPB Campfield, Cumbria

Left. Clay loom-stones or fishing-weights from the shores of Allonby and Beckfoot, Cumbria

Below left. Plaster diorama of Haig Pit at the Colliery Museum, Whitehaven, 2010

Low tide on the River Dee: buoys mark the ships' route to Kirkcudbright, Dumfries & Galloway

The RNLI All-Weather Lifeboat, the former Tyne-class *Sir John Fisher*, being lowered into the water at Workington, 2016

Above. Celtica Hav, carrying wheat, enters Silloth New Dock, where *Zapadnyy* has already discharged her molasses

Left. Holding the water in: the dock gates at the Port of Workington

High tide on a saltmarsh stint

The 'training walls' and rails opposite Glen Caple on the River Nith, Dumfries & Galloway

Sundew at Bowness Common

Cotton-grass on Drumburgh Moss

Drowned birches on the line of the former Solway Junction Railway, Bowness Common

Right. Ruthwell Cross, Ruthwell, Dumfries & Galloway

Below. Roman altars at Senhouse Museum, Maryport, Cumbria

New Red Sandstone, Fleswick Bay, St Bees, Cumbria

The remains of the viaduct on the embankment, Bowness, Cumbria

Above. Granite wall at Kippford, Dumfries & Galloway

Right. The unconformity on the shore: a 'mess of Brockram' on the Coal Measures sandstone. Barrowmouth Bay, Cumbria

Cockle shells at Kippford,
Dumfries & Galloway

Starfish on the mussel-beds of Ellison's
Scaur, off Beckfoot, Cumbria, 2005

Heston Island and disused stake-nets at Auchencairn Bay, Dumfries & Galloway

Low tide at Moricambe Bay

Scallop dredgers moored at Kirkcudbright, Dumfries & Galloway

The start of the Maryport Trawler Race, 2019, with boats from Maryport and the Isle of Man

Haaf-netters setting up the 'draw' for places in the boak, Bowness-on-Solway, Cumbria

Kirkbride airfield. Frank – wiry, slightly stooped, full of energy, and enthusiasm for everything to do with birds and other wildlife – has an almost encyclopedic knowledge of the Mosses and their history. He is also one of the small group of people who pushed forward their vision of setting up the Watchtree Nature Reserve at Great Orton after the foot-and-mouth epidemic of 2001 (see Chapter 4). Whether at his home or his office, whatever question I ask, he leaps to the bookshelves or his computer to find documents, maps, diagrams and photos. He has taken me around Watchtree and out on the Mosses on several occasions and our conversations zigzag from one enticing topic to another. For example, I learnt that on Glasson Moss, hemp or flax pools dating from 400 to 800 AD have been discovered, in which the plant stems would have been soaked or 'retted' to remove the living tissue from the fibrous stems; that although no human 'bog bodies' have been found, there is the Solway Cow – actually, pieces of two cows, which were probably ritually deposited in the bog in early medieval times; and that analysis of peat cores shows that at about a metre below the surface of Glasson Moss there is a sudden change from tree pollen to pollen from weeds and arable crops, coinciding with the Roman period, when they deforested the area to build Hadrian's Wall.

*

I've long been curious about peat cores and in April 2016 I was able to join a group of artists, poets and fiction writers, scientists of various sorts, sound-artists and composers who visited Kirkconnell Flow not only to experience what being on a Moss or Flow or mire might mean, but for a more practical purpose, too: to get our hands dirty – to go peat-coring.

Kirkconnell Flow, north-west of Kirkconnell Merse on the River Nith – along the banks of which geomorphologist David Smith and colleagues had taken cores and where, later, I would go to look for the Nith's 'training walls' (Chapter 4) – is, like the other raised mires around the Firth, a dome of peat. When the glaciers retreated they left behind a landscape that was pitted with small hollows, often lined with glacial till, a fine and fairly impermeable layer. Water accumulated and formed lochans, which were slowly filled in by fenland plants; ultimately, sediment and dead material formed waterlogged areas in which the roots of plants could no longer reach the lower layer of soil. The plants fringing these swamps used up any nutrients, so that the soggy central area became nutrient-poor – making perfect conditions for bog plants and mosses like *Sphagnum* to thrive. These mosses form a living 'skin' on top of the bog and as they grow upwards, dead vegetation slowly accumulates and is compressed beneath them, producing peat.

As the peat layer grows and thickens it becomes isolated from groundwater, and is watered only by the rain. In other words, the bog is 'ombrotrophic': a new word, one of many that I learnt that afternoon from Dr Lauren Parry, of Glasgow University, who had come to lead us through the heather to the central dome. An ombrotrophic, literally 'cloud-fed', bog is one that is wetted only by rain, not by becks or burns, and is isolated from the surrounding landscape. Rain is poor in nutrients, so only those plants that tolerate acid, low-nutrient conditions live there: the richly coloured *Sphagnum* mosses; the berry-bearing plants such as blaeberry and crowberry; bog myrtle and bog rosemary with their crushed-leaf scents; pale 'reindeer moss' lichen; and the white fluffy flags of the bog-cottons. It was early in the year when we visited, but the hints of past and future were there – even wizened last-season crowberries amongst the budding cotton. To reach the centre of the peatland we had walked a woodland path ringing with the descending chromatic scales of competing willow warblers, and waded through knee-deep tangles of heather. For decades, humans had battered and chipped away at the mire, first by cutting peat, then by burning heather for grouse-moors. Birch carr and pine forest encroached as the water table dropped and the peat dried out; the trees themselves transpired water and hastened the drying and destabilisation of the mire.

Lauren's PhD research had related to the composition of peat but she now concentrates on carbon sequestration – the ways in which peatlands store carbon and how the carbon levels relate to changes in the climate. While very well practised in the technique of peat-coring, she was exploring less 'dirty' techniques such as ground-penetrating radar to get information about the bulk density, the relative amount of solid, organic material within the peat. Until then I hadn't realised that a peat bed can be riddled with 'pipes' and cavities and crevasses. And now, out on the Flow proper, there were more words too, hard-edged words and musical words with a cadence: acrotelm and catotelm, the lagg fen and the haplotelm, the 'Russian' corer, proxies, testate amoebae.

'We really love taking cores and looking at them under a microscope,' Lauren said. 'The peat's a wonderful archive.' Why 'an archive'? Vegetation growing on the thin, living surface of the bog (the acrotelm) responds in different ways to the environmental conditions – the amount of rain, warmth, nutrients, pollution. Different species – of plants and of microscopic animals – grow more slowly or out-compete others depending on their favourite conditions. Pollen grains, even radioactive particles (remember Chernobyl), blown in from surrounding areas are deposited or incorporated amongst the stems. And as plants die and are replaced, they are grown over by new individuals. The combination of new plant growth and water-logging means that oxygen diffuses into the lower layers so slowly that the decomposition of dead plants uses it faster than it can be supplied, and so the lower layers (the catotelm) are anaerobic. Layers of only partly decomposed plant matter build up in an active mire.

Dr Richard Lindsay, of the Sustainable Research Unit, University of East London, writes: 'Peatbogs are responsive systems with homoeostatic mechanisms that are not far removed from those found in living organisms . . . features having many similarities to tree rings can be found in the equally thin layers of peat which are successively deposited in a bog over millennia. These narrow bands of peat tell the same tale as tree rings but over a much longer time-scale.'[1] And the peat is 'a direct product of the vegetation which created it'. So the peat's characteristics reflect the nature of the vegetation that created it, while the vegetation itself reflects the amount of water and types of nutrients prevailing while it lived.

So, an archive: which can be discovered and examined by drilling down and taking a core, slices of which can undergo a battery of chemical, spectroscopic, radiological and microscopic examination in the lab for its range of minerals and isotopes, for types of pollen, for the hard cases of the unicellular testate amoebae and silica-shelled diatoms, and more. Amoebae, pollen – these are 'proxies', different species flourishing or dying out according to prevailing conditions. All these measurements provide clues as to what climatic conditions prevailed at the time each 'peat ring' was laid down.

Taking a core doesn't need a drilling rig, just patience, organisation and some brute force. The day I went out with Dr Parry and the rest of the group, it required the help of 'the Russian' – which turned out to be a type of metal corer, a side-filling sampler with a sharpened edge; when rotated clockwise the blade cuts a core which is then held in the chamber by a plate. Back on the surface, the corer is rotated in the opposite direction and the undisturbed sample is slid out into a plastic gutter; the whole is then wrapped in clingfilm and kept horizontal. The first core, from the acrotelm, was fibrous with roots and decaying grass and other plants. Fifty-centimetre lengths were gradually added as the Russian penetrated deeper; each time it was then pulled back to disgorge its sample. The core from 150–200 centimetres represented life about 2,000 years ago. *Gloooop-gloop-gloop.* The corer made wet, sucking sounds as it was pulled free. Deeper cores became smoother and sloppier. More rods were added: taller, able-bodied men were co-opted to help lift the rod vertically into place. An artist was taking photos of cranberries; someone was watching skylarks through his binoculars; others were sitting on the ground chatting; someone was making notes in a tiny ring-bound book. Three of us bounced on the mire's surface several metres away from the group, and startled faces turned towards us as the quake spread out in waves. And then, at 6.5 metres' depth, the bottom of Kirkconnell Flow was reached! The sample chamber of the corer revealed a glossy grey cylinder of clay, formed by the friction of glaciers against rocks and deposited 8,000 to 9,000 years ago.

On the other side of the Firth, cores show similar depths, and more. The raised mires are dotted around both sides of the Upper Solway, some in better condition than others. Several, like Lochar Moss, were considerably larger than they are now; Bowness and Glasson, neighbours separated by a road, are probably continuous at a deeper level. Peat for fuel, but mainly for horticultural use, has been the driver of destruction. Nutberry Moss, on the right side of the A75 as you drive west to Dumfries, is only recognisable as a (former) peatbog by the mountains of fine black powder that await weighing and bagging, and although famous as a battleground between the English and Scots in 1542, the Solway Moss to the north-east of Gretna now barely exists, partly because of peat harvesting and partly due to a massive 'bog-burst' in November 1771. Back then, after 'three days' rain of unusual violence', the skin of the bog was unable to hold more water and ruptured. Thomas Pennant visited it a year later and heard the story: 'About three hundred acres of the moss were thus discharged, and above four hundred of land covered; the houses either overthrown or filled to their roofs; and the hedges overwhelmed; but providentially not a human life lost, several cattle were lost . . . The case of a cow is so singular as to deserve mention. She was the only one out of eight, in the same cow-house, that was saved, after having stood sixty hours up to the neck in mud and water: when she was relieved, she did not refuse to eat, but would not taste water: nor could even look without shewing manifest signs of horror.'[2]

The English South Solway Mosses National Nature Reserve comprises Wedholme Flow and, closer to the coast, Glasson Moss, Drumburgh Moss and Bowness Common. The first two are owned by Natural England, Drumburgh (pronounced *Drum*-bruff) by Cumbria Wildlife Trust and most of Bowness by Natural England and the RSPB, and all are in varying states of restoration. They are also a European Special Area of Conservation and Sites of Special Scientific Interest; their plants and animals are multiply protected. This wasn't always so. Until the 1500s the Cistercian monks of Holme Cultram Abbey near Silloth used peat from Wedholme Flow to evaporate seawater for salt production. People who lived around the edges of the Mosses often held turbary rights, allowing them to hand cut peat for fuel, and on Wedholme Flow, hand cutting and then machine cutting was a commercial business until fairly recently. Seen from a gyroplane or drone, Wedholme Flow near Wigton is a complex of rectangular cells that were drained and commercially harvested, before its recent re-wetting and the start of restoration.

*

The Cumberland Moss Litter Company at Kirkbride, part of Wedholme, was started by a Dutchman, Henry Engelen. Peat cutting and drying on a commercial scale requires well-founded knowledge of cutting and draining the landscape, and the Dutch had been designing and over-seeing commercial peat works in England and Scotland since the 1920s.

Patrick McGoldrick, with a shock of white hair and a soft Scottish accent, told me about his father: 'Dad worked on at least four mosses in Scotland . . . they were all owned by the Dutch. Some got worked out, they were not as deep as Kirkbride [Moss]. At Kirkbride you could dip your big ash pole in as far as you could and still couldn't feel the bottom.' Hearing that there was work to be had at Kirkbride, Patrick's father moved down to Cumberland, bringing his fourteen-year-old son with him.

I first met Patrick McGoldrick and his sister at Port Carlisle, where they were taking part in the oral history project, 'Remembering the Solway',[3] and later I visited him at their home to hear more of his stories about working on the Moss.

'I helped Dad. As life went on, after eight weeks, the school board man came looking for us. "Why's that laddie not at school?" But by then it was tattie-picking time – we had a week off school for that – so I was able to stay on. Dad went on cutting and I got a job at a farm at Kirkbride. Every time I could I went to the Moss with my dad.'

Stooling or stooking, walling, stacking; stickers, bats and spades: the language for the stages of cutting and drying the peats varies from country to country and region to region. The shapes of the tools vary, too, as do the patterns of piling peats to dry. Tools were personal and precious.

My dad used to look after his tools, he had a big stone and a file to keep them really sharp. We always carried our tools over our shoulders. Then we went back to Ireland for a week's holiday and he put them in a ditch that had running water in it, to keep the handles tight. When we came back the tools were missing. He said, 'Patrick, where did I put my tools? I was sure it was in that ditch but they're no' there. Somebody's pinched them.' There was this big guy, broad, Big Stan the Pole, and he wasn't a grafter, his tools weren't good. Big Stan would have fought anybody. And Dad said, 'Aye, he's cutting peat, I'm going to see *his* tools.' And there was the stripping spade, the tools, all with my dad's initials on them – he'd carved them on the handles. Stan was in the ditch working and my dad says, 'That's a good set of tools you've got here. They're mine.' 'No. No, Frank, you bugger off.' And Dad picked up the spade and put it to Stan's throat – he got his tools back! Anyway, Stan got some more tools. And then I saw them, sitting on the bank, and my dad was sharpening his tools for him!

'Your tools – the blades were never rusty,' Thomas Holden told me. 'You stuck them in the peat at the end of the day, and they came out really clean and shiny, it must have been the acid in the peat.' Thomas had got in touch with me when he heard that I was interested in talking to Kirkbride peat cutters, and invited me to his home in Aspatria; he had started cutting peat when he was in his thirties, and later had driven the workers' bus, and he still had happy memories, even though 'it was hard work, but I was younger then. You got worked-in!'

Patrick explained how to use the various tools. 'You put down a line then cut a mark along the line with the sticker [Dutch *stikker*], it's like a hay-knife. It's flat with a handle, and very very sharp, about fifteen inches long. Then you stripped off the heather, with like a turfing spade.' He still has his father's spade and went to fetch it from the shed. The shaft is curved, the blade pointed; he showed me how he'd cleaned it and repainted the metal part green, then demonstrated in the sitting room how to push it in at an angle and lift the turf. The layer of heather and grass stripped off the top was thrown down into the ditch 'with the heather up so it widnae die back', to keep the ditch drier and protect the underlying peat. 'The bat was used for cutting the peats, its blade was only four to five inches wide—' (he drew a little diagram on my note pad – it looks a bit like a cricket bat, with its long, narrow blade). Standing in the cut, 'You cut in at an angle, then lifted out the peat and heaved it up onto the bank.'

Peat is wet, about 95 per cent water, 5 per cent solid material; if its integrity is damaged, all that water will escape. Peat cuttings, then, were wet and slippery places, and ditches had to be dug at the ends of the banks to allow the water to flow away. Thomas Holden told me, 'It was surprising how wet it was. You've got boards on your feet, you walk like a duck!' He laughed and stood up to show me, walking bent-kneed, bow-legged, across his sitting room.

'The deeper you cut the wetter it got,' Patrick explained. 'We wore footplanks to stop you sinking. You had a piece of wood with the corners cut off. Then the bottom of a boot nailed on, and you put your wellies in and laced it up – we'd get bits of leather to strap the boot in, see?'

The area for commercial cutting at Kirkbride was mostly inferior brown peat, not good for burning, but sold to improve garden soil. Individuals and families had stints with rights of turbary around the edge of the Moss. According to Thomas, 'Some of the peat was like butter, but it was more fibrous where we were cutting cos it was used for gardens. There was peat being cut for fires too – that was black peat. It was dark and thick, and it curled up like a banana when it was dry. They used a different tool, like a bat but with a blade on the side so it was L-shaped, it could cut the bottom and the side at once.' This blade with a flange would have been like the fenland 'beckett', Irish 'slane' or Scottish 'flauchter' or the *tairsgeir* from the Outer Hebrides.[4]

The Hebridean Isle of Lewis is a mosaic of gneiss, that glittering grey and ancient rock, and lochans and runnels of water, and blanket bog. Under a grey sky, the colours are muted and merge into each other, a spectrum of greys and browns and greens, colours for which even artists probably have no names. Then, as the sky clears, the palette abruptly changes, picking out reflections on blue water, the scarlet shimmer of sundew, and the emerald fluff of the 'drowned kitten' moss, *Sphagnum cuspidatum*. And where there is light there are shadows, accentuating the rectangular depressions of peat cuttings.

In response to the first proposal for a massive wind farm and its infrastructure back in 2008, artist Anne Campbell, with Finlay Macleod and others living on the Isle of Lewis, gathered together a glossary of Gaelic words relating to the moorlands of Lewis and especially to peat cutting, which was later published in a pamphlet called *Rathad an Isein*.[5] (The book's English title, *The Bird's Road*, means the narrow gap left on top of the bank between the cut edge and the *gàrradh*, the stook or stool.) There was no copy in the bookshop in Stornoway, the capital of Lewis and Harris, so I tried the library; as I chatted to the very helpful librarians the mother-and-toddler group were circling their hands as they sang 'The wheels on the bus', in Gaelic. The library didn't have the book either, but fortunately Anne Campbell has a website and I was able to contact her and obtain a copy. From *The Bird's Road* I discovered that all parts of the *tairsgeir* have their own name, so important are the tools for the cutting of the peats on Lewis.

Another person who contacted me about the Kirkbride peat cutting was Malcolm Wilson. He had been a model-builder for most of his life, building model yachts, barques and ships like the *Bounty* from scratch, making and assembling all the parts himself. Some of his beautifully detailed models have even been sold at Christie's. But in the 1980s, the manager at Fisons – at the time the company was cutting peat commercially at Kirkbride – asked him to make models of the tools and barrows as gifts for members of staff who were retiring. Malcolm had gone to the Kirkbride workings to measure the tools, and in the course of a few years, he had made six or seven scale models. As the warm spring sunshine flooded into his front porch, he unrolled the scale drawings that he had made in 1989 and showed me photos of the finished pieces – the long, low barrow with long handles and no sides, the footboards and bats. Suddenly, movingly, the reality of the tools that Patrick and Thomas had been describing became very clear. And at the end of our meeting, Mr Wilson gave me his precious drawings and photos: I shall always treasure them.

Patrick told me about the three stages of piling up the peats, different architectures to let the air circulate and help the blocks to dry: stooking, walling and then stacking. When the peats in the walls were dry, then 'it was time to stack. We had two barrers, they were long with wooden wheels, the legs were short, mebbe six inches high . . . Dad would go way up the bunker [bank] – he'd run his barrer up and he'd build a tower to start about twelve peats high. Then you'd build the stack up against it, you'd empty your barrer and fling it all into the middle. The peats were then put in singly, up to ten feet high. You put standards, peat standing up on end, then a whole row right the way round again. The stack would be like that—' Patrick held his hands up at an angle, 'to let the rain run off.' When fully dry the peats were taken to the mill to be ground – in wagons, or bogies, drawn by a small, narrow-gauge locomotive. This meant that temporary rails had to be laid along the banks beside the stacks. 'The only time we'd get to drive the locomotives was when we were putting the rails down! . . . The rails were tied to the sleepers by fishplates and transported on a flat bogie, in pieces four or five yards long, piled five to eight high on the bogies. Two men could lift them and put them in place. They were laid down all this way and that, it was hard to move the cart. I lost a few bogies in my time!'

It was unremitting hard work, but what comes over strongly is how much Thomas, and Patrick and his father, enjoyed being out on the moss. 'On the moss, we'd get really dry, we used to run out of drink in the hot weather.' Thomas said. 'We'd gaa across the fields and gaa to the farm and get bottles of lemonade – they made it there. We was absolutely gasping . . . You could start at any time of day, some of them took their wives – it doesn't matter, as long as you get the peats.' In contrast, 'If it was really cold you could make a fire. If it was gey windy, where you'd stacked the peats you could take peats out and make an alcove, like. The peats would burn away faster than . . . they were that dry and light.' Or, if you were near the mill, there were old railway carriages and a bus, as Patrick recalled. 'Beside the mill was an old bus which was a bothy for the workers. We'd eat our pieces in there – we called it our "piece" [that's the Glaswegian term], and the men from Aspatria and Wigton, they called it their "bait".'

'And see, you'd see all the wildlife. Cuckoos – my dad would never disturb cuckoos. There were dragonflies, and adders there too . . . The only thing my dad would grumble about would be the midges – but he'd have his pipe and he'd puff the smoke. Even in the rain it was good.'

For Thomas, 'Mind you, I enjoyed it, there was a lot of freedom, there was nobody chasing you, like. What I liked about it as well, there was always a lot of birds there. There were cuckoos in summer, nesting in the meadow pipits' nests – they nested in the stacks, and there were wagtails nesting there too. And I remember another thing, all the snakes – adders. One fella said he'd chopped more snakes in two than he'd hardly cut peat. The adders'd be curled up on the path, and they slithered away when you came.'

Interestingly, Brian Blake records that when he visited Wedholme Flow he actually met Henry Engelen, who not only told him about the adders, 'but showed me one that had been killed a few days previously'.[6]

Patrick remembers his father as a hard-working, kind and honest man: the photo in the sitting room shows a man with a broad, kindly face. 'He was only a small wee man – he had hands like leather. He was a grafter. As soon as he finished with his piece he started again. I always remember him, working away with his pipe in his mouth. He'd put it on top of a peat and forget where it was. Then later he'd be saying "You know, Pat, I've covered that pipe of mine again." In stacking the peats in the springtime he'd find his pipe again and he'd always say "That's the best smoke I've had for a month!".'

Henry Engelen, meanwhile, went to Germany to look at automated cutting; he imported German cutting machines which could cut, chop, lift and lay, so that the hard labour was done away with. Engelen eventually sold the Cumberland Moss Litter Company to Fisons. After a management buy-out in 1994 Fisons became Levingtons; then Scotts took over the peat cutting and milling in 1997, until peat harvesting ended in about 2000. The Moss was skimmed and 'harrowed', the heather all gone, the peat exposed and drying out.

I spoke to Henry Engelen's son, who still lives in the county, and he told me that Fisons had given one of the locomotives, named *The Henry Engelen*, to the Mining Museum at Threlkeld. I went to look for it so that I could send him a photo, but although several men, young and old, who lovingly repaired and reconstructed old locos were there, sadly none had any knowledge of that particular engine. I was, however, shown other engines from the Mosses – but all were in various stages of decay, patiently awaiting repair. As for Henry Engelen, after he retired 'he liked his music, and his ballroom dancing – he'd often go dancing five times a week!' his son told me. He bought the old British Legion hall at Silloth because it had a good wooden floor, which apparently was perfect for dancing, and he took the floor away and installed it in St Cuthbert's Hall at Wigton.

We appreciate, now (even though this often means re-learning, and reminding others), how unique and important the Mosses are in terms of their vegetation, their animals – birds, vertebrates ranging from lizards to roe deer, and invertebrates such as flatworms, beetles and dragonflies – and their importance as a carbon store and record of the past. The peat workers, too, appreciated the wildlife and specialness of the place, but in those days climate change and carbon stores were not part of the vocabulary. 'Natural capital' means different things at different times: in Patrick's time, peat was there to be harvested and sold, and peatbogs provided work. Patrick took his father back to the Kirkbride Moss on a couple of occasions, but they were upset at the changes, the lack of heather and vegetation, the lack of birds.

*

On a humid July day a few years ago, my husband John and I left our bikes in the green, shady lonning that marked the south-eastern corner of Bowness Common and walked along the track that divided the boggy, re-wetted Moss from heathery woodland carr. In search of longer, wider views across the Firth and towards the Lakeland fells, we were heading for Rogersceugh Farm, built on the whaleback of a drumlin, thirty metres above the Moss. Drumlins are elongated hillocks of boulders, sand and gravel that were deposited beneath glaciers, and there are swarms of drumlins all around the Solway. Hollingworth's 1931 map shows an anti-clockwise swirl on the English side, reflecting the north-westerly movement of the ice from the Lake District, and Rogersceugh, a particularly spectacular example shaped a little like a cuttlefish, sits upon this pattern.[7]

Flies buzzed in clouds, clegs lurked and settled; there were wild raspberries to eat, and splashes of scarlet robin's pincushion parasitising the wild rose stems. And then a chance encounter, of the sort that I've come to expect and hope for around the Firth: men out on the heather, heaving a thin rod out of the peat and upwards to the sky. Peat corers. They had just reached the clay, they said, at five and a half metres; they were surveying the depth of peat each side of the track, because the route of the track might be changed. 'It's a bad day for clegs,' one of them said. Further on, cattle stood in the shade of trees each side of the track. A red car appeared, reversed and stopped. A small, white-haired woman got out and walked to a fence, then stood talking to the beasts who came towards her. She smiled at us as we came closer. 'Clegs are bad. You've got bare legs.' She introduced herself as Dot Harrison, former tenant of the farm, who had 'heard the cattle blaring' and had driven over to see what was wrong. Had we seen the waterlily pond? she asked. There were red and yellow and white flowers. 'The farm belanged til Lord Lonsdale so mebbe he had them planted. I did wonder if the ducks had fetched them in . . .' We mentioned the 'dismantled railway' featured on the OS map. 'It's right here, look,' she turned round and pointed at a gated track behind us, now overgrown and barely distinguishable as the former line. 'And you see the concrete there?' – blocks almost buried in nettles and brambles – 'there was a hut there for the railway. That concrete was made to last!'

We wondered about going across the field to look for the rest of the line, but the cattle were fractious, the ground was boggy, the clegs were bad; it would have to wait for another day. 'Aren't you scared to walk over the Moss?' she asked. 'I would never do it! When I lived here there were adders, hundreds of adders. I was scared stiff o' them. When the men came to dig the drains they used to go out in their lunchbreak and catch them.' As we walked past the derelict and decaying farm, the hot air vibrated with a threatening, rumbling mutter. At each side of another lonning, out of sight but within smell and sound of each other, were two bulls arguing over who was dominant; seeing us, one turned to stare and blared.

The Solway Junction Railway (SJR) was a grandiose Victorian infrastructure project, a railway driven across a peatbog and then over the Solway to Scotland. What lay behind this vision and daring disregard for topography? I wanted to find out more about the effects on the Moss and on the Firth, and was fortunate to be awarded a small grant to research it, with the help of local photographer James Smith. As well as using his usual cameras, James was beta-testing Unmanned Aerial Vehicles, or drones. Our detailed findings about the SJR and viaduct – the construction, the economics, the damage and dismantling – can be found in our separate *Crossing the Moss* project; there are many amusing and some horrifying stories, and several puzzles about the men and the materials. But the effect of the SJR on the peat of Bowness Common or Moss, and on the Firth itself, is very relevant here.

The raised mire of Bowness Moss or Common is separated from the Solway by a strip of farmland and raised beach, along which runs the narrow coast road; on the road's seaward margin the dense gorse thickets, which blaze with colour for most of the year ('When the gorse is not in flower, kissing's not in season'), grade into saltmarsh, and saltmarsh grades into mudflat. At Herd Hill Scaur between the Moss and the village of Bowness-on-Solway, a stub of sandstone juts out across the saltmarsh and the mud. Topped with grass and scrub, its sloping edges of dressed blocks are zoned with colour: grey and yellow lichen; faded rose stone that is darker red where it has been gashed by erosion; and the green handkerchiefs of sea lettuce, *Ulva lactuca*. At the embankment's free end, a row of six cast-iron pillars provides a frame for the view across to Annan, to the unvegetated embankment on the far side.

The Bowness embankment is joined to Bowness Moss by a now scarcely visible line, for the railway across the peatland was dismantled in 1934. But imagine the scene in the summer of 1869:

Heat shimmers above the domed mass of moorland; the white puffs of cotton-grass hang limply amongst spiked yellow asphodel and the sticky crimson spoons of sundews; grasshoppers chirrup intermittently, and a dragonfly buzzes a lizard drowsing on the warm peat. Half a mile to the north, the two ends of the new viaduct are striding on thin black legs towards each other, across the sea that glimmers between the English and Scottish shores. The whistle of a steam locomotive, and continuous clanging and hammering from the bridge, shake the still air. Here on the Moss, the sweating navvies work silently, almost in unison, cutting into the peat, lifting the heavy, dripping spadefuls onto the banks. They wade in murky brown water, often skidding on the smooth cut surfaces, the wooden boards on their boots squelching at each step. Lads chivvy horses that pull waggons, laden with sleepers and faggots, on temporary and uneven rails, and as the ditches lengthen each side of the future permanent way, water drains into them from the peat, and flows towards the Firth like rivers.[8]

John Brogden and his son Alex had been involved in the building of railway lines since 1838. In the late 1840s, the company had also diversified into iron ore mining in the Furness area. Furness was not connected with the mainline railways so the haematite had to be shipped out by sea, an inconvenient and expensive means of transport. It was from this basis – their expertise in setting up railway and mining companies – that Brogdens decided to link Furness to the main line via a new railway, the Ulverston & Lancaster Railway (U&LR). Their engineer on the U&LR had been James Brunlees and his railway had famously crossed, by means of two elegant viaducts, the wide estuaries of the Kent and the Leven that flowed into the unstable reaches of Morecambe Bay. Meanwhile, the mining of haematite in West Cumberland was increasing year on year, the red ore being sent to the smelting works in the south of Scotland either by ship or by railway up to Carlisle and then via the westward dog-leg through Gretna to Lanarkshire. It is little wonder therefore that Brunlees and the Brogdens discussed shortening that journey by building a new railway and a long viaduct that would cross the Solway Firth itself. They had the expertise – and they perhaps liked the idea of another big, showy project that would not only showcase the grandeur of their vision, but would also bring in considerable revenue by capturing the market for the transport of 'Cumberland ore'. Certainly such a project, pitting human endeavour against that 'boisterous obstructor', the sea, attracted great interest:

In the long list of victories over natural obstacles which have been achieved by railway promoters since Chat Moss was compelled to forget its waywardness, and afford a firm foundation for one of the first iron pathways in the world, few can aspire to occupy a more important place than the recent successful enterprises of carrying railways across tidal estuaries or arms of the sea . . . When an arm of the sea surged in protest against the further advance of a projected line, a long and tedious detour was preferred to a contest at close quarters with the boisterous obstructor. No such pusillanimous spirit, however, can now be said to exist in the frame of a railway projector . . . The boldness of the design of forming, by means of a bridge across the Solway, a new connection between England and Scotland at the most dangerous part of the Borders, as well as the difficulties which the locality offers to operations of the kind in the capricious floods and the shifting sand, and the rapid currents of the Firth, have caused the progress of the works to be regarded with more than ordinary attention.[9]

The 'Cutting the Sod' ceremony took place in Annan on Tuesday, 26 March 1865, where a public holiday had been declared, 'a great day for Annan, and Annan did its best to make it memorable'. The procession, the crowds, the *déjeuner* and the multiple speeches, are recorded minutely – including the interjected 'hear, hears' and laughter – in the *Carlisle Journal* for Friday, 31 March 1865. But it was not until September 1869 that freight trains were finally permitted to travel along the line across the Moss, and the Board of Trade Inspector delayed the transport of passengers until August 1870.

Bowness Moss was the problem. In the archive collections at Carlisle I had unrolled the eight-foot long plan of the proposed railway line across the Moss, pinning it down on the long desk with cloth-covered weights. The line of the track was deceptively simple, from a new bridge over the River Wampool to Whitrigg then north, straight across the Moss to the Rogersceugh drumlin and then on to Herd Hill at Bowness. The plan also showed the elevations of the track and the depth of the peat. At its deepest, the peat on Bowness Moss was 'fifty feet deep', a figure that accords well with modern corings that have shown depths between six and fifteen metres. Undisturbed, undrained peat doesn't form a solid substratum, but is mostly water with only about 5 per cent solid matter, and the peat bed usually contains pipes and holes and hollows. Yet nearly forty years previously Chat Moss had been conquered by George Stephenson during the building of the London & Manchester Railway; the track had been 'floated' across the Moss on wooden faggots and hurdles and ashes. Solutions were available. But although in June 1868 Alex Brogden laid the final two girders on the Solway viaduct, work out on the Moss had completely ceased and was not restarted until February 1869. The *Railway News* reported that 'Very heavy and extensive draining operations [were] being required, and infinite labour [was] being consumed in laying the way' over the Moss. There are repeated advertisements in the local papers, asking for navvies and stonemasons.

At the Cutting the Sod ceremony, navvies had been an important part of the parade. The *Carlisle Journal* reported on the 'silver spade and silver-mounted mahogany barrow for the work of the day, borne shoulder-height by four navvies in the smock frocks, red neck-cloths, and white nightcaps of the order', one of whom handed the spade to and reportedly joked with the local dignitary who ceremonially cut the turf.[10] There are several mysteries about the navvies, especially about where they lived – there is no mention or trace of an encampment. When they were working on the Moss they would have been eating and drinking enormous amounts of food and drink, for navvies had to be strong, fit men – agricultural labourers who joined up could not, at first, stand the pace. Joseph Firbanks (contractor for many railways, including the Settle & Carlisle) 'mentions quite casually that his navvies consumed on average two pounds of meat, two pounds of bread, and five quarts of ale a day'.[11] There were deer, hares and wildfowl on the marshes and Mosses of the Upper Solway, fish in the rivers and the Firth, crabs and brown shrimps among the rocky scaurs – it's highly likely that the navvies supplemented their diets by scavenging and poaching. But by the end of 1869, all the equipment was being sold by auction in Annan, and the navvies had moved on. We know almost nothing about them and their living conditions, or their views about Bowness Moss.

Their work was lengthy and laborious but basically, two ditches were dug parallel to the proposed railway track, each at a distance of one chain (sixty-six feet, or twenty-two yards) and linked by cross-ditches. Water flowed out of the cut edges of the peat and poured into the River Wampool to the south and the Solway to the north, and the damaged peat sank by four to five feet each side of the track. (We now know that the integrity of peat is disturbed at least a hundred metres each side of a cut, so that water 'bleeds out' through the lesion.) Two layers of faggots (bundles of wood) were put down to spread the load and on top of these was laid a ballast of broken sandstone and clinker, then the sleepers, to which were attached the rails. In some places even this proved insufficient and double-length 18-foot sleepers had to be laid over the worst parts to spread the load even further, covered with more layers of faggots, upon which were laid the usual sleepers. Ninety thousand faggots were laid in total!

The *Whitehaven News* reported that on 26 June 1869 the SJR 'which is just completed', was tested by a 'party consisting of the directors, engineer, and contractors of the line, with two of the largest and most powerful engines belonging to the Caledonian Company'. The *Cumberland Paquet* of 6 July 1869 was a little more cavalier with the truth (not quite 'fake news', but it makes an entertaining tale):

THE FIRST RAILWAY ENGINE ON BOWNESS MOSS

The Solway Junction viaduct across the Frith did not prove the most difficult part of the undertaking in an engineering point of view. The crossing of Bowness Moss was a much more perplexing undertaking. After the deep drains had been in operation, and *the water had been running in river-like streams on each side of the proposed line for many weeks* [my italics] and the faggots in thousands had been laid, it was expected that something like a good foundation had been obtained. This hope was delusive. A line of rails was laid, and an engine was run along, but one fine morning the engine nearly disappeared. It sank down into the moss, and looked like bidding farewell to the scene of its labours; but fortunately for the contractors the steam was up, and the engine had strength enough to drag itself out of its perilous position. Now all difficulties appear to have been surmounted.

Unfortunately, several 'difficulties' followed, of which the worst was the severe and almost terminal wrecking of the viaduct by ice floes in January 1881 (see Chapter 1). The viaduct had previously been eulogised by the *Whitehaven News:* 'Its appearance is light and elegant, and in the full tide it will have the semblance of a piece of enchanted workmanship resting on the bosom of the racing waters.' But after the ice floes hit, the rails swayed in the wind above the 300-yard-long 'Cumberland gap' and the 50-yard gap on the Scottish side. Yet the damage was eventually repaired and the SJR struggled on, becoming less and less viable economically as cheaper iron ore was imported from Spain. It was finally closed in 1926, and the railway and viaduct dismantled in 1934.

*

In December 2016 John and I decided to explore as much of the remaining railway track at the Bowness end as possible, hoping we could at least reach the cutting through the small drumlin, the 'mineral island', in the middle of the Moss. As we squelched and jumped across tussock and mire to the birches and willow that marked the line of the track, two roe deer raised their heads above the scrub, their large ears cupped towards us and then, white rumps flashing, bounded across the bog and into the fringing woodland. At this northern end, close to the former station (now a private house), we found a series of embankments that were probably remains of the sidings where cattle were once loaded into wagons. The ditches were still running with water and fringed with the skidmarks of leaping deer, but the line of the track was hard to see amongst the trees. Out on the Moss again we followed trods through the tussocks and heather and through patches of sodden dark brown peat. Two buzzards circled silently in the distance, spiralling higher and higher until we lost them. At this northern end of the Moss were also traces of former peat cuts, shallow straight-edged banks, now almost hidden in the heather.

Sun shone from a pale blue sky and the air was crisp, the moorland around us a palette of wintry browns and the blonde and red moor grasses. There was no sound apart from the occasional peeping and trilling of oystercatchers on the distant shore. We could now see the railway embankment on our right and we headed towards a more obvious patch of raised ground that was bordered by bracken. There was a stretch that wasn't overgrown, but was pale and grassy-green. To reach it we had to leap across a broad ditch that ran parallel with the embankment, and then a few metres further on a short section of a

cross-ditch, at right-angles and almost hidden in the long grass. I prod-
ded the surface of the embankment with my walking pole – and it
struck something hard at once. Scraping away the grass with our bare
hands, we found clinker and broken pieces of dressed sandstone. Ballast!
I never imagined that one day I would be so excited to find clinker and
ash. John paced out the distance from the line of ballast to the ditch: it
was about twenty-two paces – one chain. Pacing in the other direction, he found the other ditch. One was almost dry, the other contained
slowly flowing water.

We now followed the track into the line of trees, and into a shallow
cutting. This was the mineral island, where the navvies had escaped
from the peatbog and had cut through firmer, drier glacial till, leaving
pebbly banks on each side. An elongated pit, now partly filled with
water, suggested a 'borrow-pit' for obtaining stone, perhaps used to
ballast the track across the bog. The trees – willows, birches, the occasional holly – self-seeded from as long as seventy years ago, and tangled
with thick trunks of old-man's-beard, made an obstacle course of our
explorations. Everywhere was still and quiet as we sat on a mossy
branch to eat our lunch; no sounds of birds or the wind, nothing but
the ghosts of a vanished railway, and its navvies. I thought I might
return in the spring when surely this strip of woodland and scrub –
like other hedgerows and copses – would be bursting with the songs
of chiffchaffs, willow warblers and chaffinches. I tried to imagine
what it had been like in the past: a single passenger coach linked to
wagons loaded with iron ore and cattle, pulled slowly by a puffing
engine across a seemingly empty and, to many, slightly frightening
landscape, across ground that had barely been 'tamed', raw brown peat
glistening each side of the track, and ditches with dark running water.
Perhaps, if you lowered the window, you might have heard the
bubbling call of a curlew above the rhythmic rattle of wheels on rails.
The smoke would briefly be constrained by the sides of a cutting,
smuts billowing against your face and clothes, before the moorland
opened out again, and you would see the gleam of the Solway not
far ahead.

Even after the railway was dismantled, the peatbog continued to drain in those 'river-like streams' into the ditches, its surface drying out and the vegetation changing, its stored carbon being oxidised into the atmosphere. Then Natural England and the RSPB set to work to restore the Moss. Both organisations and the now-retired Frank Mawby shared their inspirational work with my collaborator James and me for our *Crossing the Moss* research: maps, core-sampling data, photos and, best of all, their time and enthusiasm. Bowness Common includes the RSPB's Campfield Reserve, and both the pastureland around the edges and on the Rogersceugh drumlin now have a system of drains and bunds and sluices whereby the water level is managed; in the winter months the level is raised further to provide wetland conditions for waders and the great flocks of grazing barnacle and pinkfooted geese, so characteristic of the Solway over the autumn and winter, that come flighting south to escape the Arctic freeze. Where there were areas of bare, cracked peat, white-beaked sedge is forming dense mats; scrubby birch is dying and rotting as the water level drowns its roots; the heathland heather, *Calluna*, is being replaced by cross-leaved heath, *Erica*, typical of wetland areas. Three species of insectivorous sundew, with their scarlet, sticky leaves; bog myrtle; bog rosemary; and even low-growing cranberry are present. Spikes of bog asphodel, their flowers bright yellow in the summer and seed heads a rusty brown throughout the winter, form carpets. Bog asphodel is known locally as 'bone-breaker': like many wetland plants its tissues are low in calcium ions, and it seems to have had to take the blame for the broken bones of poorly nourished sheep that were put to graze on the moorland. Lizards scurrying along the boardwalk at Bowness, large hairy caterpillars and empty brown cocoons of oak eggar moths, frogs and dark brown toads: there's plenty of animal life to observe. I once saw a single adder on the dry heath and a few years later, two at nearby Drumburgh Moss: one of them was coiled in a pool of sunlight, sleeping, but he must have felt the tremor of my footsteps for he opened his eyes and raised his head to consider my presence, then slipped away without fuss, as softly as a bale of silk unwinding. Wherever there is wetland

and vegetation, there will be dragonflies and damselflies too: a bird-watcher I met described them as the 'birds of the insect world'. At Bowness Moss I saw nine different species in a couple of hours, including the enormous emperor dragonfly, *Anax imperator*. There are real birds too, reed buntings, larks and pipits, even the occasional red kite or osprey from across the Firth. As palaeontologist and naturalist Richard Fortey says, when writing about the current inhabitants of his piece of woodland on the South Downs, 'A list of animals and fungi could become tiresome, but it is necessary to grasp the true richness of nature. Think of it as not so much an inventory as a catalogue leading to compelling and interacting stories . . .'[12]

Out on the Moss I try to stop 'list-ticking' species, and to take time to imagine how their 'compelling and interacting stories' might play: to imagine their stories in three dimensions – burrow into the ancient peat, bask in the sun on a boardwalk, hide amongst *Sphagnum* floating in a pool, flit above the heather, rise up into the air. And then throw in the fourth dimension, of time: day by day, week by week, through the seasons . . . the years of growth past and future; the smells, of wetness and hot dry heather; and the sounds, especially of the past. This requires some effort and certainly some patience, but I want to imprint on my mind the importance of the Moss – of *any* Moss or peatland.

As for the damaging effects of the railway, there are now blocked drains, bunds and – most dramatic of all – dams across the longitudinal drainage ditches, to hold back the flow and raise the water table. The birches that colonised where once there were rails are already pale skeletons, their branches bereft of twigs and leaves, their feet drowned in the small lakes that reflect the sky. This is what the dying forest each side of the Firth must have looked like, when the sea moved in (Chapter 2). Perhaps one day – remember peat accumulates at the rate of ten centimetres per century – these trees too will be trapped and preserved in peat.

Restoration of all the South Solway Mosses is underway, their return to wetlands helped by the increase in the *Sphagnum* mosses – as many as fifteen different species on Bowness Common. *Sphagnum* has the unusual ability to take up more than ten times its weight in water, and its metabolism ensures that its surroundings become weakly acidic. It grows upwards, its lower parts dying as it grows and, because of the acidic and anaerobic environment, this layer of dead material does not decompose, but becomes compressed, eventually forming peat. Carbon is sequestered: carbon dioxide that has been taken out of the air by growing plants and incorporated into molecules such as carbohydrates and proteins remains tied up in the undecomposed material, and is 'locked up' or stored.

There is little to see of the former railway: the concrete remains of a hut at Rogersceugh Crossing, some broken sandstone ballast, the indistinct lines of the former ditches. But ferreting around near the overbridge by the former Bowness Station I found a stone tunnel through which water is still trickling out onto the saltmarsh. Its narrow wavering channel across the mudflat is a reminder of a time when drainage of the peatlands was considered necessary and useful.

6

Red

Mr Goodchild and Mr Idle had found accommodation at the Ship Hotel in Allonby, and Goodchild had been on a recce.

> Of course there was a reading-room. Where? Where! why, over there. Where was over there? Why, *there*! Let Mr. Idle carry his eye to that bit of waste ground above high-water mark, where the rank grass and loose stones were most in a litter; and he would see a sort of long, ruinous brick loft, next door to a ruinous brick out-house, which loft had a ladder outside, to get up by. That was the reading-room, and if Mr. Idle didn't like the idea of a weaver's shuttle throbbing under a reading-room, that was his look out.[1]

'Goodchild' and 'Idle' were Charles Dickens and Wilkie Collins. In September 1857, they were being guided up Carrock Fell, in the Caldbeck Fells to the north-east of Allonby, when Mr Idle hurt his ankle: he 'was within one step of the opposite bank, when his foot slipped on a wet stone, his weak ankle gave a twist outwards, a hot, rending, tearing pain ran through it at the same moment, and down [he] fell'. We know the story and its sequel because they wrote about themselves, as Goodchild and Idle, and their adventures, in *The Lazy Tour of Two Idle Apprentices*. Collins was taken, via Wigton, to an unnamed village, where he was treated by a doctor. After a day or two of rest, and boredom, 'Goodchild' then 'converted Thomas Idle to a

157

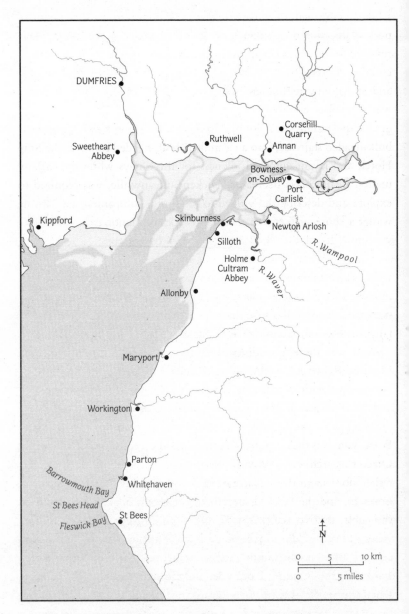

Sandstones

scheme he formed, of conveying the said Thomas to the sea-coast, and putting his injured leg under a stream of salt-water . . . he immediately referred to the county-map, and ardently discovered that the most delicious piece of sea-coast . . . was Allonby on the coast of Cumberland . . . and at Allonby itself there was every idle luxury (no doubt) that a watering-place could offer to the heart of idle man.' They set out with great expectations. At Allonby, Collins hauled himself up 'a clean little bulk-headed staircase, into a clean little bulk-headed room' at the Ship Hotel, and lay in enforced idleness for three days while the inflammation in his ankle decreased. Dickens, meanwhile, was required to explore and describe the village to his sessile companion: perhaps he wasn't a shore-walker, for their story doesn't mention the shore. He was, though, unjustifiably rude about Allonby village.

In fact Allonby was already quite a well-known watering-place, with a large colonnaded bathhouse where the water was heated for the more sensitive bathers, and several little cobbled courtyards and fine stone houses – but of course that doesn't make such a good story, especially for such a metropolitan as Dickens!

'And was there a reading-room?' Idle/Collins asks. A red-brick Reading Room had indeed been established above the weaving shed, but this was soon to change. Thomas Richardson, a Quaker, banker, and local philanthropist, commissioned the construction of a new building with considerable financial backing from his cousin Joseph Pease, who was the country's first Quaker MP. Pease appointed another Quaker as architect, Alfred Waterhouse (who went on to design the rather more magnificent Natural History Museum in South Kensington, London), and the tall and substantial brick building, held in trust for the public by five Trustees, was opened on 28 July 1862 – nearly five years after the Idle Apprentices' visit. During the mid-twentieth century, though, the building fell out of use. The Trustees sold it in the 1970s and the Reading Room decayed, its disintegration accelerating when the roof collapsed during a storm. But happily, in 2005, it was bought by a family who gradually restored it and turned it into a private dwelling; its red-brick walls are now crisp and bright.

If Dickens had indeed strolled along the Allonby shore, he could not have helped but notice the discoid pebbles of red sandstone lying on the multi-coloured shingle. If he had collected one to take back to Collins in his 'little bulk-headed room', they might have observed its pleasingly rounded and symmetrical outline, the slight roughness of the sandstone that sparkled with mica, the heft of it in the hand that suggested a perfect paperweight.

At about the same time that they were stumbling around Cumberland, a man with a 'beaming ruddy face' and a 'sunny nature' was attempting to piece together the geology both sides of the Solway. Robert Harkness became interested in geology while studying at Edinburgh University; in 1848, aged thirty-two, he moved to Dumfries and his research on the geology and fossils of south-western Scotland soon brought him to notice so that in 1853 he was appointed Professor of Geology at Queen's College, Cork, and very quickly became a member of the scientific establishment, being appointed a Fellow of the Royal Society (London) and of the Royal Society of Edinburgh. His professorial duties, according to his obituary, 'did not deprive geology of his active labours in the field; he simply added new explorations to his former areas, and we find him at work . . . "On the Silurian Rocks of Cumberland and Westmoreland" [and] "The Permians of the North-West of England"'.[2] It was those Permo-Triassic rocks – especially the New Red sandstones of Dumfriesshire and the Cumberland coast and Eden Valley – that interested Harkness.[3] He was well acquainted with the New Red of the Eden Valley, because he frequently visited his sister in Penrith.

As you travel around the Solway coasts of Cumbria and the eastern parts of Dumfries & Galloway, the dominant colour is red. There is red dustiness in the areas where haematite was mined, the ore from which iron is extracted; haematite mines were found along the Cumbrian coast and down to Millom, where the poet Norman Nicholson lived throughout his life. The transport of this iron ore was the driving force behind the building of many of the railways, including that of the doomed Solway Junction Railway and the viaduct across the Firth.

Red

There is, too, the redness of brickwork that crops up in surprising places – like Allonby's Reading Room – where one might have expected the native red stone to have been used. But bricks were cheap and easy to make, and the red clay was easily mined. The predominant redness, though, is found in the molehills, walls, houses, and many of the large Victorian municipal buildings – the libraries and town halls, especially on the Scottish side – that are built of New Red Sandstone. There are different shades of red, from palest pink to dark liver, and the colour is often hidden by plates and stipples of yellow or grey lichen, but nevertheless it is the sandstone, the country rock, that influences the character of the area. Because so much stone has been needed for building, there are also dozens, if not hundreds, of sandstone quarries scattered around the landscape. Many are derelict and almost hidden beneath brambles and scrubby moss-hung trees, but others glow with the freshness of recently quarried or half-dressed freestone. Many of the Solway's stories are grounded in its sandstones.

*

A few years after I came to live here, I was very fortunate to join a patrol on the Firth by the then fisheries protection vessel, *Solway Protector*, which was berthed in Whitehaven marina. We waited in the sea lock as the water level dropped nearly three metres before the great curved gates could be opened to let us out. The outer harbour was still calm, but as we passed the sea wall and the boat accelerated away, the bow rose up, the engines growled, and we ploughed noisily into the waves, which became more and more confused as we rounded St Bees Head. Spray cascaded against the windscreens to be swept away at once by the wipers. At fifteen knots we left a wide trail of churned green water, and I hung on tightly as the boat crashed into the waves. We passed a raft of guillemots, which took off and skimmed swiftly across the white wave-tops; a gannet beat slowly overhead, twisting its head to look down at the surface of the sea. The dark red cliffs of St Bees Head were splashed with white guano from the nesting guillemots and fulmars, and to the south the fells were dark, the top of Skiddaw hidden. In the changing light, the outlines and contours of the distant fells were as pliant as sandcastles washed by the sea. The Upper Solway is described by horizontals, soft lines of mud and marsh and the Mosses, but here, and across the Firth around Rascarrel, there is the hardness and verticality of cliffs. And on both sides there is that pervasive redness, in drystone walls, houses and bridges and quays.

I had recently bought a sculpture, 'Phoenix in Flames', chiselled out of St Bees sandstone by Sky Higgins, a young sculptor from Maryport, and this, and the view from the sea of the eponymous cliffs, made me want to find out more about this stone. So David Kelly took me to Sandwith, north of St Bees, for a personal 'geology tutorial'. He had had plenty of experience in leading geology field trips in this area, both for his sixth-formers and for members of the Cumberland Geological Society. He is tall, white-haired and relaxed, and he strode apparently effortlessly up and down the cliff paths throughout the morning; he was also full of enthusiasm (and patience), and knew exactly where to find all the best features. We walked along a narrow path northwards from Birkhams Quarry where the recently quarried blocks of sandstone were still a raw, bright red, and – because the sandstone layer is tilted – we also walked backwards through time, to a face where the blocky layers of rock were whitened by lichen, and partly hidden by fern and heather. But despite the vegetation there were plenty of fascinating clues to find, and we scrambled around on the banks to look at the exposed rocks. The layers of sandstone were interspersed with shallow bands of red shale, like flaked tuna between slices of solid rye bread. The shale, formerly silt that had been compressed and solidified, was easily fragmented, soft and shiny when rubbed with a finger. This sandstone complex – the rock and shale – was laid down in flash floods and braided rivers. And this kept on happening, for millions of years: the floods bringing down silt and ground rock, quartz and fragments of granite, from a large mountainous mass to the south. The mountains were ground down and carried away in pieces, eventually forming new rock. Robert Harkness had speculated that nearly five miles' thickness of rocks had been eroded between the oldest rocks of the central Lakes and the limestones and coal measures around the edge. The sediments were continually compressed, until the sandstone layer – the St Bees sandstone formation – was one kilometre thick: new rock from old. As always with the Solway landscape, it is a thrill to discover that parts of it are so unimaginably ancient – and have gone through such unimaginable changes. The greasy red mud smeared on my fingertip and notebook was 230 million years old, give or take ten million years or so.

Naming of names: there's something exclusive but poetic, even dramatic, about scientific jargon: 'load casts', 'shale flake conglomerate', and small pockets of muddy 'rip-up clasts'. There was evidence everywhere to show that this sandstone was sedimented from rivers. A rock face was marked with parallel lines, indicative of fast, strong currents; on one block the lines folded back on themselves, suggesting the soft sediment was folded, perhaps when a river bank collapsed. Here and at nearby Fleswick Bay (where later I could use my newly acquired knowledge) there was obvious 'cross-bedding', sloping lines that intersected the horizontal, signs of ancient sandbanks intersecting with variously angled slopes. There were boulders with bulbous growths, where the sand was pressed down into mud before it was hardened into rock. Elsewhere meandering ridges on the surface of the block showed where the Permo-Triassic mud dried out and fresh sand spilled into the deep desiccation cracks.

Not all New Red sandstones are the same. David handed me a fragment of the rock, turning it to catch the light. It sparkled. 'You see those glittery bits? They're mica flakes – they settled out of the water. If this sandstone had been wind-blown, like the Penrith sandstone, you wouldn't see any mica because the particles would have been well sorted on the dunes. The grain in St Bees stone is less than a half-millimetre in diameter. You can't pick out the grains under a lens.'

'Beestone', as it's known in the building trade, is 'a fine-grained stone, dull red in colour . . . it is susceptible to weathering in very salty environments . . . The abrasion resistance is towards the lower end of the range.'[4] Mica flakes and quartz add glitter. Sculptor Sky Higgins, expressing a quiet, almost shy enthusiasm for her work, had told me about the sparkling. 'When you're carving,' she said, 'the stone's covered in dust and just looks pale. The warmth and the hidden grain come through when you wash it. There's a lovely sparkle, you can see the tiny specks glistening in the light. The sandstone's incredibly abrasive on tools – it's sandpaper! I use the files and then I take a piece of sandstone – something I've cut off the block, a sandstone pebble, anything, to sand the surface. And then for the fine sanding I use wet and dry paper.' Finally the piece has to be washed and polished, and a little oil rubbed in. 'Fine sandstone is quite silky, you get a crisp image, the maximum sculptural effect. With sandstone there's no reflection of light to distort what you see.' The rock's colour, beautifully brought out by polishing and oiling, is due to the presence of haematite, in which the iron has been oxidised to the red ferric form; in the St Bees formation, the sediment was exposed to oxygen in the rivers and on the floodplain. This is in complete contrast to the red Penrith 'Lazonby' sandstone, for that was formed in arid desert conditions, as high, wind-blown dunes. The quartz grain inclusions are large (referred to as 'millet seed') and often look frosted and abraded under the microscope. The iron in the Penrith sandstone was oxidised in the hot open air.

Sadly, Sky Higgins is no longer sculpting, but there are natural sculptures to enjoy down on the shore. One November three of us launch ourselves into the gale ('It's invigorating!', 'Blows away the cobwebs!', and so on) to walk across the fields above those high red cliffs of the St Bees headland. Below us shags are paragliding straight-winged, leaping from the ledges and swooping downwind before looping back to do it all again; one after another after another, mesmerisingly repetitious. We slither down a slimy track strewn with wind-blown plastic and wrack, through a narrow entrance, ahead of us the V of shore and sea – and then walk out into the vast amphitheatre of Fleswick Bay, an echo-chamber reverberating with the sound of waves and gulls. Pebbles roll like ball-bearings beneath our feet, their colours gleaming where they are wet. Ragged clumps of foam are being blown along the shore, torn from the glistening white mounds that fringe the incoming tide.

Foam clings to our hands and boots and I wish my grandchildren were here to scoop it up and cover themselves with iridescent bubbles.

Monumental rounded and undercut figures and abstract shapes huddle beneath the cliffs, and the platform that reaches out into the sea is patterned with lithified ripples and deep pools pink-lined with living tufts of *Corallina*. Coloured pebbles, sand grains, shadows and the changeable light that ricochets between sky and sea challenge perspective and perception. The sculpting by waves and water-stirred stones has created delicate hollows and ridges in the glowing stone. Norman Nicholson understood so well the forms and textures of the stones of Cumberland:

> The stone is grained,
> Smooth as walnut turned on a lathe,
> Or hollowed in clefts and collars where the pebbles
> Shake up and down like marbles in a bottle.
> Here the chiselling edges of the waves
> Scoop long fluted grooves, and here the spray
> Pits and pocks the blocks like rain on snow.
> Slowly the rock un-knows itself.[5]

Those 'chiselling edges': Sky Higgins knew about the fragility of sandstone. 'Sandstone won't stand up to being carved too fine or too thin. It can break – and if it does, you know it's broken for a reason, because it was too weak or you tried to do something it wasn't suited to. But you learn how to work it, how to use it, you get a feel for what you can and can't do. I build up a bond with the piece, I get a feel for the stone.'

On the shore at Fleswick I notice how all three of us keep stopping to stroke the contours of the rock. My geologist husband points out cross-bedding and load casts and other manifestations of those names that David Kelly taught me. The friend who is with us is an industrial archaeologist, an expert on quarrying; he has become distracted by marks high on the cliff near the entrance to the bay and he climbs up to have a closer look. The rock face is scarred by the marks of pick-axes – men must have worked there to cut and remove some blocks of stone. But how would they have transported them – by horse and cart? The blocks must have been small because the entry to the shore is steep and narrow, scoured out by the tiny beck. It is more likely that the stone was man-handled to a waiting boat at high tide.

There are much easier places to acquire red sandstone, like Birkham's Quarry on top of the cliffs just to the north. Records show that the site has been quarried since 1861; it might even have been worked prior to this. The quarry is part of an area of many conservation acronyms and designations and is about 800 metres from the RSPB's reserve for the cliff-nesting sea birds.[6] So when a coastal footpath was rerouted to skirt round the seaward edge of the quarry there was an outcry; however, from the new path, now well blended with its surroundings, are excellent views: into the quarry itself, outwards along the coast to the north-east, and across the Firth to the cliffs near Auchencairn. You might see seals or porpoises, or even basking sharks. Birkham's is a small and historic quarry, part of West Cumbria's and the Solway's cultural heritage. Recently the owners were granted a licence to continue extracting stone, partly on the grounds that this type of sandstone is important for restoration and renovation projects, not only here in the UK but also overseas.

Fleswick Bay is the sole cove on the English side of the Solway and is a place where smugglers have brought their goods ashore, where boats have been wrecked and where families have come to enjoy the sea. For decades, people have carved their names along the cliffs: families from the nearby mining village of Kells are particularly well represented. Some names are part-eroded, others hidden by green slime, some look as though they had been scratched on a lazy after-noon – but one stands out: *Judy McKay*, engraved deeply and with handsome serifs, the superscript 'c' underlined. The oldest that we found was 'M.I. 1774', almost hidden by algae.

*

In the year that M.I. carved his initials, Henry Duncan was born at Lochrutton in Kirkcudbrightshire. His father was the Minister of Lochrutton Church, and Henry eventually followed him into the ministry. In 1828, now as the Reverend Dr Henry Duncan, Minister of the Church of Scotland in Ruthwell, Dumfriesshire, he presented a paper, 'An Account of the Tracks and Footmarks of Animals found impressed on Sandstone in the Quarry of Corncockle Muir, in Dumfriesshire', to the Royal Society of Edinburgh. He had been given a slab of red sandstone from the quarry, across which were fossilised footprints, and he had sent casts of the footprints to William Buckland, who was by then President of the Geological Society and a Canon of Christchurch, Oxford (he was later appointed Dean of Westminster, with a living in Islip, near Oxford). Buckland was a palaeontologist and geologist, famous for his lively lectures and eccentricities – he reputedly 'ate his way through the animal kingdom' – and he thought the footprints might be of reptiles, since they would have been the only animals that could have made such tracks on land at the time when the sandstone was laid down. The Bucklands kept an odd menagerie of animals in their house and garden, including several reptiles, and one of the footnotes to Henry Duncan's paper to the Royal Society of Edinburgh includes a letter to him from Buckland:

Oxford, 12th Dec. 1827. *1st*, I made a crocodile walk over soft pye-crust, and took impressions of his feet, which shew decidedly that your sandstone foot-marks are *not* crocodiles. *2d*, I made tortoises, of three distinct species, travel over pye-crust, and wet sand and soft clay; and the result is, I have little or no doubt that it is to animals of this genus that your impressions on the new red sandstone must be referred, though I cannot identify them with any of the living species on which I made my experiments . . . I conceive your wild tortoises of the red sandstone age would move with more activity and speed, and leave more distant impressions, from a more rapid and more equable style of march, than my dull torpid prisoners on the present earth in this to them unnatural climate.[7]

John Murray III, of the publishing dynasty, 'sought and obtained permission to make a circuitous journey home, in order that he might inspect some remarkable prehistoric foot-prints of animals in the sandstone at Ruthwell in Dumfriesshire, and afterwards made his first acquaintance with the English Lakes. These same foot-prints caused no little stir among the geologists of the day', according to John Murray IV's biography of his father, which also includes this entertaining account:

Jan, 23, 1828. I went on Saturday last to a party at Mr Murchison's house, assembled to behold tortoises in the act of walking upon dough. Prof. Buckland acted as master of the ceremonies . . . At first the beasts took it into their heads to be refractory and to stand still. Hereupon the ingenuity of the professor was called forth in order to make them move. This he endeavoured to do by applying sundry flips with his fingers upon their tails; deil a bit however would they stir; and no wonder, for on endeavouring to take them up it was found that they had stuck so fast to the pie-crust as only to be removed with half a pound of dough sticking to each foot. This being the case it was found necessary to employ a rolling pin, and to knead the paste afresh; nor did geological fingers disdain the culinary offices. It was really a glorious scene to behold all the philosophers, flour-besmeared, working away with tucked-up sleeves. Their exertions, I am happy to say, were at length crowned with success; a proper consistency of paste was attained, and the animals walked over the course in a very satisfactory manner; insomuch that many who came to scoff returned rather better disposed towards believing.[8]

Buckland's ideas about tortoises are no longer accepted, but the footprints are still considered to be reptilian. Part of Henry Duncan's collection of fossilised footprints from Corncockle Quarry is on show at the museum at Dumfries. The impressions of the heel and toes of *Chelichnus duncani* are clearly seen; nearby are four-toed imprints of another reptilian (incorrectly labelled *Loxodactylus*) from Locharbriggs

Quarry near Dumfries. Next to them, in a low glass case, and open to show a coloured engraving of the magnificent Corncockle Quarry, is the rare book, *The Ichnology of Annandale*, by Sir William Jardine, 7th Baronet of Applegarth, on whose land Corncockle stood. Jardine (1800–74), a natural historian and ornithologist, was a friend of Professor Robert Harkness, and together they set about a study of the quarries and the footprints; at Corncockle the tracks are 'more common and in general more perfect than in any of the other quarries', and 'animals different in form traversed the sand'. Jardine comments that *Chelichnus duncani* footprints extend '31 feet in length . . . not keeping a straight course but zig-zag and winding . . .'[9]

Henry Duncan (1774–1846) has other legacies apart from footprints. As you drive along the A75 from Gretna towards Dumfries and not long after the sign for the Devil's Porridge Museum, you come to a sign on the left for the Savings Bank Museum. When I was young I had a red metal book-shaped 'Savings Bank' into which coins could be inserted and only retrieved with a key, but savings banks are not at all the same as piggy banks. Through speeches, pamphlets and an Act of Parliament, the Reverend Dr Duncan was instrumental in opening the first savings bank in the country in 1810, into which people could pay their money and receive interest on their savings; this eventually became the Trustee Savings Bank.

Duncan also rebuilt the broken Anglo-Saxon Ruthwell Cross, an eighth-century red sandstone 'preaching cross' on which there are high-relief images, and runes and Latin words. It is an extraordinary work, standing about 5.5 metres high, with its foot in a specially sunk depression so as to accommodate its height inside the attractively simple, white-painted, Parish Church of Ruthwell. The two main faces are carved with scenes from Christ's life – the preaching of John the Baptist, the Visitation and Annunciation, Mary Magdalen washing Christ's feet with her hair – hands, faces and garments vividly and skilfully carved in the pale gritty sandstone. On the narrower sides are intricately carved vines, amongst which perch birds and lizards and other animals.

The Cross's story is of subterfuge, love and labour. It is unknown who carved it with such love and skill. But it was the Reverend Gavin Young, minister of Ruthwell, who employed subterfuge: the General Assembly in 1640 had ordered that 'idolatrous monuments' – anything with a popish cross and images of the Virgin and Christ and saints – should be 'thrown down' and destroyed. The Reverend Young rightly perceived that the Ruthwell Cross was an important and ancient monument and, according to the local guide book, was 'firmly convinced that no effort should be spared to save it from the fate with which it was threatened'. But he needed to keep his job and manse because he had rather a large family – the words on the stones that cover their graves imply that he and his wife Jean had had twenty-eight children – so he ordered a trench to be dug in the floor of the church, and the shaft of the cross to be lowered into it and covered. It was not until the church needed to be re-floored in 1780 that the pieces of the cross were lifted and placed in the churchyard. After Dr Henry Duncan was ordained as Ruthwell's minister in 1799, he gradually gathered together the various pieces – St John the Baptist and his Lamb were discovered during the digging of a deep grave – and rebuilt the cross, adding newly carved pieces in a slightly darker sandstone to replace the missing cross-beam. He didn't dare to set it up within the grounds of the church, but there was nothing to stop him raising it in the garden of his manse next door; and there it remained until his successor, the Reverend James McFarlan, had it restored to the church in 1887.

The crosses of Ruthwell and Bewcastle (which is in England, about thirty miles east of Ruthwell) have been described by Nikolaus Pevsner as 'the greatest achievement of their date in the whole of Europe . . . The technical mastery [of Bewcastle] is as amazing as at Ruthwell. How can it have been possible, in stone, and at so early a date?'[10]

*

It was another of those delightful Solway coincidences. Judy McKay phoned me because her daughter had seen an article I had written about the engraved names at Fleswick Bay. She was no longer called McKay (it rhymes with 'tay' not 'tie') – that was her maiden name – but when she began to tell me the stories of how four generations of her family had been involved with quarrying and stonemasonry, I was hooked. On my subsequent visits to her house she showed me photographs, letters, written accounts and newspaper cuttings.

Judith's great-great-grandfather and his family came from Caithness and settled near Lazonby in the Eden valley, and he continued with his previous occupation, quarrying, near Dalston, Carlisle. Quarrying carried on into the next generation, too; one son, James, Judith's great-grandfather, owned Shawk Quarry near Thursby where sandstone had been quarried since Roman times. In 1875, James visited Philadelphia and, having seen Brown Connecticut stone on sale, thought it might be possible to sell red sandstone from Shawk there too. At that time sailing barques regularly brought grain to Whitehaven, Maryport and Silloth from Philadelphia and other American ports, and these ships often needed ballast – usually in the form of gravel – to stabilise them on the journey home. The more I discover about the Solway, the more I discover these tendrils that reach out from a person or animal or place and wrap themselves around one another. Mud and gravel along the shores: *Corophium* perhaps being scooped up too into the hold of ships (Chapter 1) and carried in wet ballast to the New World.

In June 1879 an Italian barque called the *Eliza C* brought a cargo of wheat into Maryport, and James arranged for the ship to carry out blocks of Shawk red sandstone instead of gravel as its ballast. His son Tom was ordered to try to sell the stone in America. Tom left on a ship from Liverpool, and arrived long before the *Eliza C*, which had become becalmed. When the vessel eventually arrived he found the stone was liable for excise duty of about a dollar a ton – he had to cable home for the large sum of about £100 to be sent to cover it. On top of this setback, the local architects and builders were not keen to buy the Shawk stone because it was fine-grained and they thought it would deteriorate too rapidly in their climate. Tom finally managed to get rid of the consignment by selling it cheaply to a firm in Newark, New

York, although this meant reshipping it in barges. The whole venture made a serious loss and his father James would have nothing more to do with America. But Tom was clearly undeterred and struck a deal with another New York firm, Sherwoods, and with a quarrying firm in Annan on the Solway coast – and soon Sherwoods 'sent orders for thousands of tons', Tom wrote later.[11] Seeing the lucrative American market opening up, Tom became a stone merchant, and 'took all the blocks that Cousins of Whitehaven could give me from the Sandwith quarry [above St Bees], also from Doloughans, Bankend Quarries, and all the blocks from Hy. Graves Aspatria quarries and also my father's Curthwaite quarries'; there was stone from Annan quarries too. Judith's great-grandfather James's quarrying and contracting business was expanding meanwhile, and becoming very important in Cumberland and further afield. His sandstone and his masons were required on many important projects, amongst which was the Maryport & Carlisle Railway, including all the bridges. The McKays' quarrying business expanded west to Aspatria and south to the Sandwith and Bankend quarries near St Bees.

Between 1939 and 1951 Judith's grandfather, Tom, also took over the Camerton Coal and Brickworks, near Workington, where red fire-clay bricks were made and, during World War II, supplied to the Royal Naval Armaments Depot nearby. The brickworks were associated with the Camerton colliery because not only coal but fireclay was mined in the drifts; this association was common in the small West Cumberland collieries at the time, and today there are people who scour the shores to find bricks stamped with the names of the different works.

As for Tom's sons, one of them – James – became a stonemason, serving his time first over at Annan and then down in Barrow, working on the Furness Abbey Hotel. Later he moved to St Bees. 'I said, "Oh Dad, you didn't go out with her? She was a horrible woman!"' Judith laughed as she told me how she had discovered the words 'James McKay loves . . .' cut into the red sandstone of the cave in Fleswick Bay. 'He said, "Well, she wasn't so bad when she was young!".' Later, though, James carved the name of 'D K McKay' – Dorothy

Kathleen – in a more prominent place, for this was the woman he married, and who became the mother of Judith Mary McKay. Judith remembered going to Fleswick Bay twice with her father, taking sandwiches and a screwtop bottle of lemonade. 'He'd wrap up his tools in a cloth and take them with him, and when we got there he'd say, "You sit there now and be a good girl".' She didn't remember him carving her own name, but she did remember riding on his shoulders as he carried her back home, up the hill.

*

With the coming of the railways came also the possibility of even wider distribution of stone, both along the shores of the Firth and much further afield. There is a strong likelihood that stone from Corsehill Quarry near Annan was used to build the Scottish embankment for the viaduct of the Solway Junction Railway. The Caledonian Railway Company opened a railway connecting Glasgow and Carlisle in 1847, and it was not long before branch lines connected the Corsehill Quarry to the main railway system. Temporary rails would have allowed connection with the Solway Junction Railway as it grew on the Scottish side. But the origin of the red sandstone used for the viaduct embankment on the English, Bowness, side is not clear. The Maryport & Carlisle Railway built by James McKay's company could have been used as transport for part of the way (perhaps even for sandstone from the McKays' own quarries), but the rail link across the boggy Moss wasn't finished until *after* the embankments and viaduct had been completed. Perhaps the sandstone for the English side came by sea? After all, steam barges were working in the viaduct area, towed to and fro depending on the height and state of the tides by the wooden paddle-steamer, *Arabian*. Records show that much of the Corsehill stone was indeed taken around the head of the Firth by rail and then by the Carlisle & Silloth Railway to Silloth. The port of Silloth was built in 1859, with railway sidings and hydraulic cranes, and Corsehill stone was shipped to Belfast, Liverpool and London. From Silloth, on a decent tide, it is no distance to Bowness: Scottish New Red or Cumberland New Red, the stone could well have arrived at the English embankment of the developing Solway viaduct on the high tide. Or, more simply, barges might have brought the stone across from Annan.

There's a nice story about Corsehill stone in James Irving Hawkins' book about the sandstones of Dumfriesshire.[12] The Capitol building in New Albany, New York State, was completed in 1896 and has three grand staircases, all of Corsehill sandstone. He tells us that the Great Western Staircase (later known as the Million-Dollar Staircase due to its cost) is described in the guidebook as a 'celebration of the stone-carver's art', for there are more than 400 carvings of the faces of a mixture of famous and 'everyday' people and relatives of the masons. During this period, Corsehill was exporting more than 250 tons of stone a week to their New York agents, and there are records of stone-masons from Lochmaben, Brydekirk and Templand working their summers in America. As Hawkins says, 'It would be nice to think that some Annandale great-grandmothers have their faces carved on the Million-Dollar Staircase beside Lincoln and other early presidents of the United States!'

A connection between the Solway sandstone quarries and Albany continued, but this time using the New Red of St Bees. In 2011 I visited Blockstone's quarry and works near Penrith; they were working on blocks of Beestone, cutting and preparing them to be sent to Albany for the restoration of the Cathedral of the Immaculate Conception. A man with ear defenders was using a screeching saw to cut a groove; long, intricately shaped pieces, faintly patterned with lines of deposition, were lined up on pallets; two vicious-toothed circular-saw blades, each a couple of metres in diameter, were propped against a wall. On a board in the peace of the office were pinned photos of Albany's cathedral and an architect's drawing of the façade and the spire and rose window, which showed the size and shape for every stone that needed cutting.

*

There are so many sandstone quarries, large and small, bordering both sides of the Solway. But when you see what the Victorians constructed with the stone that number comes as no surprise. Take Parton, just north of Whitehaven, for example. In the late seventeenth century and early eighteenth century Parton was a busy and thriving little town. There were two tanneries, a glassworks, a saltpan, and considerable other businesses associated with coal mining. There was also a port for the export of these goods. Clues to these former endeavours still exist along the shore in the wave-sorted patches of débris among the rocks and shingle – fragments of decorated china and lumps of thick green glass scoured by the sand, the base of a bottle with 'WHITEH' in raised glassy letters, and cloudy green marbles from bottle-stoppers. My granddaughter found a piece of blue and white china, with three Chinese figures crossing a bridge, and when we returned home we looked up the story of the willow pattern, of how Koong-se eloped with her lover Chang, and why they eventually turned into doves; I was able to give her a Wedgewood willow pattern bowl that had been my parents', one edge carefully mended by my father.

Smoothed and eroded lumps of red brick abound here on Parton shore, as elsewhere on the Solway's shores, the names of the works often still visible on their surfaces. The shore also reveals more permanent reminders of the industrial past of this stretch of coast, where coal mining and steel making once steered the economy and peoples' lives. In places the sand and shingle have been washed away to expose twentieth-century artefacts, hard banks of concrete-like ash with embedded coloured, gaseous pebbles, the slag from the Bessemer converters which poured down, hot as molten lava, from tilted railway wagons at the edge of the land.

It was the Solway's wild waves that eventually finished off the port and the export business. In February 1796 a tremendous storm thrashed the west coast of Britain, and Parton's port was totally destroyed, its sandstone quay swept away, the bay's sandbanks and shingle shifted and redistributed. With the loss of the port as its hub, it seems that Parton went into a decline.

The village, hidden at the base of the great red cliffs, is unseen from the main coastal road, and the shore is hidden from the village by a high red wall that is punctuated by a few low tunnels through which a car or horse and wagon might drive, carefully, to the shoreward side. The clues to the wall's existence are high above, spiky silhouettes of gantries with signals, and warning signs: the embankment carries the coastal railway on which Parton is a request stop. Behind the village are more high sandstone walls, in which holes are inhabited by jackdaws who peer down, heads tilted, as my husband John and I walk by. Two rusty hinge-supports point to a former, now bricked-up, doorway, behind which there is the sound of rushing water. These stone walls are the artificial face of the cliff, built to restrain the soil and shale from slipping down onto the street. From a higher vantage point we look down onto the roofs of houses and into their backyards, a view that takes in the railway, the bay, and distant church and Moresby Hall, and that is still – with notable absences – recognisable from Percy Kelly's paintings and drawings. Author Chris Wadsworth records that in an interview in 1964 Kelly 'said there was an urgency in drawing and painting [Parton] because it was disappearing as he worked': the council were demolishing many of the old houses which were unfit for habitation and were relocating people to the new housing estate on top of the hill.[13] Kelly was born further up the coast, at Workington, in 1918, and lived for a while in Allonby. His extraordinary story has been told by Chris Wadsworth, who as a former gallery owner in Cockermouth has done more than anyone to bring Kelly's works to our notice. He was a difficult man – paranoid, befriended by the rich and famous, a cross-dresser who signed off his letters as 'Roberta' (long before Grayson Perry, and while living in a rural area in less tolerant times). He was a man who inspired many unflattering epithets, but whose works, including his now famous illustrated letters and envelopes, have become highly sought-after and admired. In a letter to his friend, the Millom-based poet Norman Nicholson, in June 1971, he writes of a trip he and his first wife Chris had made: 'Parton beach fascinates us both. Here we collected tons of sea coal and made friends with some of the old characters, ex-miners, who frequent the beach.'[14]

We head south along the Wagon Way, a smooth track once used by horses hauling coal towards the port and now ideal for cyclists, walkers and buggy-pushers, and we are distracted by more red walls: low walls against the cliff's base, and a waist-high wall above the railway embankment, curving around the headland into the distance. The walls demand our time, our attention; they are so perfectly constructed of blocks, some as much as a metre long and a third of a metre high. Red sandstone, St Bees stone, some of it possibly quarried from the cliff between Parton and Whitehaven, then transported to where it was needed; the block faces cut and dressed and worked to almost perfect squares and rectangles. Imagine the difficult and unsafe work that went into the quarrying and transporting of the stone, the cutting, the dressing, the lifting and the placing. How many men, how long, did it take for the walls of Parton to be built?

Where large sandstone buildings have been constructed on the flatter, soggier parts of the Solway's fringes, there is always the question of how the stone was brought there. The Cistercian abbey of Holme Cultram, to the south of Moricambe Bay, was founded in 1150 by monks from Melrose Abbey in Scotland;[15] at that time the area was still part of Scotland, and indeed Robert the Bruce's father is buried there (which didn't stop the Bruce from viciously attacking the abbey in 1319). Erected in a low-lying part of the Solway Plain on a mound of glacial till and clay known as The Holm or The Island, and bordered by wetlands and saltmarshes, it was nevertheless an imposing building of red sandstone. After the Dissolution in 1538 it became the parish church, and was gradually reduced in size after various structural accidents and as its stones were purloined and repurposed. A shockingly stupid arson attack in the summer of 2006 nearly destroyed the church but it was rebuilt and eventually reopened in 2015.

During his walks through 'Middleland', Rory Stewart, then MP for Penrith, visited the shell of Holme Cultram's church while it was still being rebuilt. At the end of each day's walk he would email his former-diplomat father to tell him what he had seen, and at five o'clock one morning his father replied 'in capital letters' with some thoughts and suggestions, 'Cistercians. They sound particularly admirable chaps . . .' But, as a rearer of sheep myself, I'm pleased that he homes in on the really important questions: 'And do we know whether their wool was special, or merely plentiful? Indeed, any idea of what sort of sheep there were at that time?'[16] It seems that the abbey may have had about 10,000 sheep, probably of various breeds since they would have been scattered around its various granges, and wool was shipped out from the abbey's port at Skinburness to places around the Irish Sea. There is a reference to the monks of Furness Abbey in South Cumbria (formerly Lancashire) wearing brown habits made from the wool of young Herdwick sheep – this would surely have felt like doing penance.

Just north of Skinburness, on Grune Point, there are reputed to be the remains of a 'Viking chapel'. Whether this is true or not is difficult to determine, just as with the tradition that the haaf-net beam is the length of a Viking oar, or that Herdwick sheep were brought over by the Vikings. The Herdwicks, with their pale grey fleeces, sturdy white legs and kindly white faces, are the iconic fell sheep of Cumbria, a tough and hardy breed accustomed to living up on the fells for most of the year. For several years I bred and raised Herdwicks on our lowland smallholding, and I have a great respect for their anarchic and intelligent behaviour. Although we have recently switched to rearing Hebridean sheep instead, I continue to use Herdwick fleeces for weaving rugs. Young lambs are born brownish-black, but by the autumn their legs are white, their faces go through a speckled phase as they turn colour, and over the next eighteen months or so the fleece itself changes to grey. That grey fleece has a short staple (fibre-length), mixed with short, tough hollow fibres known as kemp, and with its softer underlayer it makes a warm and water-repellent coat for sheep during the Cumbrian winters. The shorn fleece is perfect for rugs but very rough for clothing, and although the younger brown fleece is slightly softer, with less kemp, it would still be unpleasant to wear. Cistercian monks, however, wore white or pale habits of undyed wool, while it was the lay-brothers who wore the brown. And whether this wool actually came from Herdwicks of different ages is not at all certain because, as Geoff Brown, a former Secretary of the Herdwick Sheep Breeders' Association (HSBA), convincingly shows, it is only in the last 200 years or so that the Herdwick as we know it today – clean white face, a white leg at 'each corner', fleece changing from brown to grey – has been developed through selective breeding.[17]

As for the breed's Viking origin, a study in 2014 by Professor Dianna Bowles – herself a Herdwick breeder – Amanda Carson, Secretary of the HSBA, and geneticist Peter Isaac, of the genetics of Herdwicks and another upland breed, the Rough Fell, found that both breeds might have had a historical association with the primitive Pintail sheep of Texel Island, just north of the Netherlands. The island is known from archaeological remains to have been a major trading post for Vikings as they moved down into the North Sea and beyond. Herdwicks also share a rare gene with sheep from North Ronaldsay in Orkney and from breeds within Finland, Sweden and Iceland – so there is indeed a faint possibility that they might have arrived with the Vikings.[18] After all, the sails of Viking ships were woven from wool 'from the long, coarse outer hairs of a primitive breed of short-tailed sheep' and each sail required a lot of wool (and sheep).[19] It makes a good story.

Since 2006 the West Cumbria Archaeological Society (WCAS) and Grampus Heritage have been carrying out a series of digs to try to determine the size of the former abbey of Holme Cultram, and have uncovered cloisters, a refectory, graves, ditches and more (see 'Salt', Chapter 8). But in 2016 a dig to the south of the abbey revealed a stone jetty about a metre beneath the surface, and WCAS received a grant to carry out further exploration in May 2019. I was puzzled as to why a jetty would be located there, several kilometres distant from Moricambe Bay and, since I had met Mark Graham of Grampus previously at a conference, I walked over to his office in the village next to ours to find out more. The research materials now available online are an embarrassment of riches – LiDAR, maps ancient and modern, academic and other papers – and I and a good-natured black labrador, who had strolled out of an adjoining office, stared at the screens as Mark explained how the discovery had come about. South of The Holm there is a confluence of three waterways – the Stank Beck, the Crummock Beck and the River Waver, which is the river that enters the south of Moricambe Bay by Calvo Marsh. An Ordnance Survey map had shown a mill at this point but the building was no longer visible, so the archaeologists carried out a geophysical ground survey (the sort of survey the rest of us might be familiar with from Channel 4's *Time Team*). The result showed a dark line in which there was a break and, from other excavations, Mark suspected this break might indicate something interesting, so they started to dig – and uncovered part of a jetty. It's now clear that the channel on which the jetty was built had silted up after the Waver was diverted away from the track of the Carlisle & Silloth Bay Railway in the nineteenth century, but prior to that it would have been an important waterway for the abbey. Mark told me how he had watched the sea racing up the Waver as high tide approached, and that the water was still flowing inland where the three waters met. He has no doubt that laden barges could have waited out beyond the mudflats and the marshes, and then used the power of the tide to be swept up the channel.

This, then, must surely have been how the sandstone for the construction of the abbey arrived, especially if, as thought, it came from Scottish quarries. The channel and jetty would have been used for the transport of goods in and out – including wool – via the port at Skinburness and, after that had been swept away in a storm in 1301, the new port at Newton Arlosh. Recently, Mark and the team have found several stone clapper bridges out on the marsh and at the edge of the peatlands at Wedholme Flow. Each bridge is made of three large sandstone slabs, supported at each end. Given that the monks were involved in collecting brine on the marshes and needed to burn peat in the salt-making process (Chapter 8), it is likely the bridges belonged to them. As Mark said, 'Nobody but a monastery would invest in sandstone for bridges.'

*

Red

I first visited Barrowmouth Bay near Whitehaven during that 'geology tutorial' with David Kelly. On the cliffs at Sandwith, the shadows were finger-numbingly cold and the wet grass soaked the bottoms of our jeans. We skidded down the slick muddy path, through the brambles and stunted trees. It was lumpy, difficult terrain formed by the shales which had slipped down from the sandstone face, with deep half-hidden holes. The path that followed the tramway or inclined plane to the gypsum mine was buckled and indistinct, interrupted by the cliff's slippage. There was a small square bridge of neatly cut and faced red sandstone blocks; the remains of a sandstone pump house and weigh house, both tilted backwards by the rotated slip face so that their blank doors and windows pointed skywards; a few hard white rocks of anhydrite; and 'John Smith 1935' – whoever he was, he had carved his name at the top of the cliff, and the looped and flowing script of his name was carved here too.

We lost the path on another landslip, and when we finally reached the shore we were muddy and damp, but we were now amongst a field of boulders that had tumbled from the cliffs above us. Some had drifts of mica embedded in them, where the delicate flakes had once been swilled by the river's current into a small depression. Others had marks of ripples that were formed on the bed of the river delta when the sandy sediments were deposited; still others showed swellings or circular indentations, where stones or mud were pressed against the sediments as they were compressed and turned to stone. But neither the marine algae nor the salty white encrustation could completely hide the colour of the slabs beneath our feet. It too was sandstone, but of a deep purplish colour, a different age, a different origin. Where we were standing was once a humid tropical swamp, about 290 million years ago in the Carboniferous, when 'proto-Britain' was near the equator, stuck in the middle of the Pangaean super-continent and slowly sailing north. This was Coal Measure Sandstone, 'Whitehaven sandstone'; you can see it further up the coast too at Parton. The coal miners working far out beneath the sea would have seen a less colourful rock above them, dull and pale, because it was deposited in stagnant, anaerobic water. Only later, at the start of the Permian, when the swamps dried out and the area became arid desert, would the surface become oxidised to that characteristic purple-red.

'Come and see this! This is what I *really* like about this part of the shore,' David called. It looked like a large, shallow spill of concrete, one to two metres thick, into which gravel and chunks of rock had been stirred before it set: it looked artificial and unattractive. This unconformity and the way it arose is extraordinary. The surface of the Coal Measures Sandstone had been smoothed and hollowed, slowly eroded as it was exposed and oxidised in the desert air – and then suddenly, into every dip and joint had poured this mess of 'Brockram' or breccia, forming a dramatic contrast of colour and texture. In the Permo-Triassic a river had flash-flooded out onto the sandstone, its powerful currents carrying a mixture of volcanic material, limestones, sandstone and even Ennerdale granite from the Lake District region, and had deposited this breccia in a fan.

So we stood there on the unconformity and looked up, through tens of millions of years and a staggering range of climate change: swampy Carboniferous sandstone and a patch of alluvial Brockram beneath our feet; then about five metres of pale Dolomitic limestone formed by magnesium-rich water percolating through limestone in a shallow salty sea; muddy shales rising for an uneven, vegetated hundred metres or so; and crowning it all the great cliffs of red St Bees sandstone, deposited on an arid plain by rivers and flash floods, accumulating in a layer that was once 1,000 metres thick.

Not long ago I returned to Barrowmouth Bay with John. The line of the old tramway, up which truckloads of rock had been pulled by a steam winch, was now scarcely more than a steep green track cropped by the ubiquitous rabbits.[20] In the dark mouth of an adit from which water trickled, almost obscured by ferns and fallen stones, someone had placed two bright white lumps of gypsum. Gypsum is hydrated calcium sulphate, $CaSO_4.2H_2O$. In its finest form it is alabaster, and at the start of the nineteenth century Barrowmouth was mined first for this mineral, which was used for ornamental purposes such as monuments and for making moulds for the nearby Whitehaven Potteries. When gypsum is heated the water is driven off, and the anhydrite so formed is used to make plaster; as this type of plaster replaced lime plaster in the late nineteenth century, so the requirement for gypsum increased and the mine expanded. It eventually closed in 1908 because the contamination with naturally occurring anhydrite – which, unlike the manufactured product doesn't absorb water and 'set' – was too high. But gypsum is also a source of sulphuric acid, and in the 1950s the Marchon Company set up their chemical works at nearby Kells and mined the deep seams of the anhydrite that lay just above the coal measures. This anhydrite was used until the 1970s, when the company switched to using pure sulphur from elsewhere; the works and all the associated businesses closed in 2005 and were demolished. The mine, of course, remained, and has gradually filled with fresh water; this is relevant to a new venture, as we shall see.

Returning from the shore along what was once the winding pony track that pre-dated the tramway we noticed an unnatural shape hidden in the brambles. We excavated with a geologist's hammer and our hands, pulling away rugs of green moss and fern, and found two rectangular sandstone troughs. One large, one small, both filled in with soil and dead moss; perhaps they had been drinking troughs for the pack-horses that dragged the alabaster-laden carts. We could just decipher the letters 'KESH' carved above the trough; the same letters were also carved into the stone of the small bridge.

The last part of our route down onto the shore had had 'protection', as rock climbers say, with a knotted rope fixed to an insubstantial metal stake. Nervously entrusting ourselves to this dubious safety feature, we picked our way through the boulder field to find the purplish Coal Measures Sandstone, and the mess of Brockram 'porridge'. Then we sat on the platform of unconformity and drank coffee, watching the sea, hearing the regular *thump-thump* of an engine, a trawler unseen in the fog heading west to the Irish Sea.

Deep under our feet were the 'roads' and drifts of the Whitehaven coalfields, the mines stretching far out beneath the Firth. Coal has been mined here since the 1700s, and the last pit – Haig – closed in 1986. There are still people around here who worked at the mine, below or above the ground, and remember the noises and the dust, and – without romanticising it – the way of life and companionship. Ex-miner Tommy Norman started work in Haig Pit in 1962 and when I asked him the naïve but obvious question, 'what was it like?', he surprised me. 'It was the best days of my life! It was the companionship.' His eyes twinkled and he laughed. 'We were all related by drink! We were paid out in the pit yard or the pub, each gang was paid for the coal they'd cut and it was shared out in the pub. "One for you and one for you and one for you." And if there was coins left over, the kids got them. There was always kids hanging round the yard on Fridays.'

After the pit closed, Tommy Norman became a guide at the Haig Colliery Museum at Kells just above Whitehaven, and has many well-practised and amusing stories to tell. About fifteen years ago, and before it received a grant to 'improve' it (it has since closed), the museum was a wonderfully eclectic jumble of engines, flywheels, carts, dusty boots and lamps, where you could browse for hours and find new and often puzzling objects. I will never forget one particular exhibit which gave an almost shocking insight into the sheer scale of the collieries, the extraordinary size and interconnectedness of the three-dimensional maze of tunnels beneath the sea, just out there beyond the cliff-top pithead. It was an undated map of the collieries, hand-drawn and coloured by one Ted Wilson. His drawing showed the outline of the headland and the harbour and, stretching out under the Firth were blocks of colour, each representing a mine: green for Haig, pink for Saltom, brown for Kells. Within the blocks of colour were finely drawn and lettered parallel lines and names of the thousands of 'roads' and faces and tracks beneath the sea, mile upon mile of them, with some of the pits interlinked by roads that acted as escape routes. And if you looked at the dates, you saw that Saltom was first sunk more than two and a half centuries ago, down to 456 feet below mean sea level; in 1819 men dug it even deeper, to 778 feet – then it was closed in 1848. King Pit, sunk in 1750, closed in 1790; Kells, sunk 1737, closed 1878;

William Pit ('the most dangerous pit in the kingdom') was sunk in 1804 and closed in 1955. In contrast, the mighty Haig Pit is modern: started during World War I (hence its name), its No. 5 shaft reaches down to the Main Band at a deep 1,200 feet, and the workings were dug out nearly four miles under the sea.

Ted Wilson's hand-drawn map revealed even more: dotted across the mainland were tiny circles that indicated small, sealed-off workings, with names like Knockmorton Pit, Burnt Pit, Wood-a-green and Thicket. Most of them would have been forgotten, Tommy Norman told me – were it not for the subsidence they cause. 'All these la'al pits, the reason why people knows about them is because houses are built on top o' them!'

At Fleswick Bay we had found many names of residents of Kells carved into the red cliffs. It is likely that many of them had been miners. The late Norman Hammond, a highly knowledgeable and likeable naturalist, who knew about coastal animals from mudshrimps to minke whales, once told me that when Haig and the other pits closed for the annual holidays in July and August back in the 1960s, he would see as many as a couple of hundred miners and their families in the bay, enjoying the sea and fresh air. At that time large numbers of basking sharks used to come to St Bees Bay and Norman would drift among them in his small boat. On one occasion some of the miners shouted out to him from the beach, asking if they could come out too. 'One man said he'd love to swim with them and I told him to slip over the side – off came his clothes and in the nude he swam with the sharks from Fleswick to St Bees Bay.' For several years thereafter, Norman would pick up miners from the beach or Whitehaven harbour and take them out to see the sharks, sea birds and cetaceans. Norman, who died in 2005, was one of those rare people, so well-informed and enthusiastic about the natural history of the Solway that he wanted to share his knowledge with everyone. He was a storyteller – if you asked a simple question he would reply with such a wealth of information and anecdote that you became caught up in the tale. When he told me that one of the miners had given him a freshwater mussel that he'd found in the mine, I never forgot it.

So I immediately thought of Norman when geologist Eric Gozlan mentioned 'mussel beds' during a talk about West Cumbria Mining's plans. WCM had been exploring the still-rich coalfields that stretch out from St Bees beneath the Firth with a view to extracting metallurgical or 'coking-coal', a high-quality, low-phosphate coal which is used in the steel-making process. An enormous amount of relevant geological data already exists due to the previous decades of mining but, as Eric pointed out, there is a considerable amount of faulting in the coalfield – in other words, there have been upheavals and slippage so that a band of coal and the rocks that sandwich it might have dropped several metres relative to their main position. Understanding the depth and extent of any fault is obviously important – there's no point trying to carry on mining a band of coal if it has been pushed aside by a fault. So, to map the coal bands in three dimensions, it is necessary to drill down at different locations and extract cores. As Eric said during his talk, 'The exploration geologist is always trying to find out where he is.' And one way he (or she) can navigate down through the rocks is by looking for specific markers of geological time. That's where the fossil mussel beds become so important. Discovering my interest in the mussels from the mines, Eric had generously suggested that I come and visit their 'core store'. There, he brought out the chart from the British Geological Society's Memoir for West Cumbria, which shows the two main mussel bands, one marine, one freshwater. The Vanderbeckei Band – its local name is the Solway Band – 'extends from Moscow to America' and it is the 'target boundary' between the Middle and Lower Coal Measures that they look for in the core, identifiable by its fossil marine fauna such as the brachiopod *Lingula*. Then they search for other recognisable features. 'We crack open the cores and look for freshwater mussels. Within the next ten to fifteen metres we should then typically find Yard Band, which is a good unfaulted band.' And above the freshwater mussel band are the Main Band and Bannock Band, both coal seams that WCM will be working. Out in the core store metal shelves were stacked high with plastic core trays; black or mottled or wrapped in paper, the cylindrical cores were labelled with letters and numbers to identify their place and depth of origin. Eric gave me a section of a core that had been varnished to show the freshwater fossils more clearly: there were two very obvious bivalve shells embedded in the rock, both probably the mussel *Anthracomya*. *Carbonicola* (the only species whose name I knew) is older, found lower down. Two other sections of core show fragments of marine fossils, of the brachiopod *Lingula* and a couple of graptolites.

Seeing, *touching,* those fossils – marine fossils and freshwater fossils – *in situ* in rock that has been brought up from 400–500 metres below the seabed of the Solway, was thrilling. It's too easy to become blasé about fossils; there are boxes and boxes of them in museums and private collections. But 'my' *Anthracomya* were living, growing and filter-feeding on the bed of a river perhaps 300 million years ago. That river ran across land that had previously, on occasions separated by tens of thousands, even millions of years, been submerged beneath the sea. The river delta had itself been inundated by the sea much later, again on widely separated occasions. That competition between sea and land had been carried out as the climate had swung between tropical and temperate and glacial; the place those swamps and sediments and rocks occupied on the surface of our planet had drifted and been pushed through different latitudes and longitudes across the planet.

The piece of varnished core is sitting on my desk now, as black and shiny as the lumps of obsidian I found in Iceland. Norman Hammond would have enjoyed the story.

If it receives final approval, WCM's new mine, the Woodhouse Colliery, will stretch out under the Firth close to the lines that Haig Pit worked. The above-ground works will be situated on the site of the former Marchon chemical works, and access to the undersea works will be through Marchon's former anhydrite mine.

*

In November 2018 Tern TV's *Britain at Low Tide* crew had just spent time learning about the disaster at Whitehaven's Wellington Pit in

1910, when 136 men and boys died after an explosion in the mine. I met up with the team later and we hacked our way through the scrub along the top of the embankment at Bowness, to make the 'viaduct' part of the programme about the Solway for Channel 4. Back in July when the producer and some of the team had come to do a recce, the Firth had been looking its best in the sunny, quiet weather, and a harbour porpoise had even chosen that time to entertain us. But now we had five takes over several hours, and the wind was fierce and keen, the water choppy. By the end presenter Tori Herridge and I were cowering in a hollow between takes, swathed in scarves and hats, blowing on our hands. The producer and the camera crew had also wanted to look for traces of the railway out on the Moss, but fortunately they soon realised the impracticality of lugging their equipment across a soggy bog. However, they were inspired by the stark drama of the viaduct's six remaining pillars, and the stories told by the tumbled stones. The gently sloping sides of the embankments are constructed of partly dressed blocks, but recent storms have opened up the stony carapace to show what lies within. The centre is filled with rubble and puddled red clay. If we had visited at a very low tide we would also have seen that just offshore there are at least five more rows of stumps of the piers that held the pillars. The original plan for the embankments, according to *The Scotsman*, had been that they should reach out far enough from each side of the Firth for the viaduct to be only 800 yards long, but this plan was revised, 'fears being entertained that the current would thereby become too confined, and its force increased to a dangerous extent, the bridge was lengthened to 1700 yards'. Subsequently, it was decided to extend the viaduct by yet another 200 yards on the Bowness side, so the number of piers was increased to 193, each of five cast-iron columns – a 'light and elegant' piece of workmanship, indeed – and to close the proposed gap in the centre that would have allowed ships passage to Port Carlisle.[21]

This would not only have an effect on the scouring and deposition of sediment each side of the embankments, but there were also social, knock-on effects, on 'the once busy little bathing village of Bowness', and on 'Port Carlisle, now as lifeless and silent as any place well could be'. Port Carlisle had been in decline for a while after the closure of the Carlisle canal, and steamer trade had decreased, but now 'The viaduct of the Solway Firth has shut up the little place entirely within itself. Sea communication west of the viaduct is quite cut off except for small boats . . .'[22] Moreover, as Chris Puxley, former harbour master at the port of Silloth, wrote in the local newssheet, the *Solway Buzz,* in August 2014, 'the building of this viaduct had a profound and detrimental effect on the regular cross-Solway paddle-steamer services and day trips between Silloth and the Scottish ports of Annan, Dumfries, Carsethorn and Glencaple, with those services declining and finally ending around 1878'. But pleasingly, when the ice-damaged viaduct was closed for three years, small coastal vessels from Annan, such as the Nicholson-built schooner *Syren*, were able to slip through the gaps carrying cargoes of sandstone to Silloth port.

Artist Alison Critchlow and I wandered around Port Carlisle – a favourite place for both of us – on an April day of changing light. It is a place that Percy Kelly would have liked to paint, with its strong, straight lines: a double row of gnarled posts which once supported the steamer pier stutters out into the water, and the jagged outline of the coaling wharf. There are red sandstone bollards and speckled grey bollards of Criffel granite, some upright, some fallen sideways, and the wide, wide space of the mudflats and marsh, and the ribbon of the incoming tide.

Metre-high patches of saltmarsh have grown in front of the quay, their eroded faces striated with history and, mudlarking, I prised a piece of an elaborately patterned Victorian tile from an edge. The tide's slack period in the Upper Firth is long, so we reckoned we still had enough time to walk across the mud and gravel to the coaling wharf. On the way we found some protruding wooden stumps and so we spent time looking for the others, the remains of a pier that had joined the wharf to the shore. There had been rails along the pier, upon which horses pulled wagons filled with coal. I thought about the horses, stepping across reverberating wood with the noise or sight of the water beneath them, and although I am not enthusiastic about horses, I could imagine the whites of their eyes rolling fearfully beneath their blinkers. But the stumps demanded investigation; I knelt to feel the raised concentric growth-rings, and see how the depressions had been colonised by tiny winkles. One stump was draped with thin green strands of *Enteromorpha intestinalis*, another decorated with drying brown spiral wrack, *Fucus spiralis*, and plaques of small barnacles had colonised the edges. Larger winkles, *Littorina littorea*, squatted on the surface or had drawn grey muddy trails with their feet on the surrounding stones. The wood smelt dull and wet. But the tide would be turning, and it was necessary to move. Beneath our feet now were gaggles of mudsnails, and the burrows of mudshrimps. We talked about the animals' meanderings on the mud, the 'calligraphy of the shore'. 'Listen!' I said, and we squatted down to hear better. The mud was 'popping', the *Corophium* were ventilating their burrows, or perhaps they were communicating. There was no way to know.

Out on the coaling wharf the sandstone blocks were pocked with holes. The wharf's separation from habitation gave us new perspectives; it was like being on a wreck that was sunk in the mud and soon to be surrounded by the sea. Neatly abutted blocks, fallen blocks, stone staircases, the village bathhouse, the village quay: like the other ports along the Solway coast, Port Carlisle too is described by New Red Sandstone. As 'country rock' in the cliffs and bays, or as rock that has been quarried and cut to shape, the rusty red sandstone, speckled with pale lichen, intervenes between land and sea.

*

The ever-changing mudflats glint at low tide in silted-up Port Carlisle, and along the River Annan, and adjoining the saltmarsh each side of the Bowness embankment. James McKay of St Bees was learning his mason's trade at Annan. Cycling by way of Sandwith to pay the men at the family's quarry, then north through West Newton, past Kirkbride and the saltmarshes at Newton Arlosh, he finally reached the railway station at Bowness-on-Solway. The railway viaduct across the Solway was the quickest route to Dumfriesshire so, as always, the station master allowed James to step down onto the track and carry his bicycle across the bridge.

7

Mud life

The muddy edges of the Upper Solway might appear monochrome and monotonous but, having stared down at the mudflats from the gyro-plane, I had seen their subtle patterns and contours that trapped light and water, revealing the past and present effects of the sea and rivers. Down at shore level the differences in colours and textures and observable life seemed very obvious. I have learnt that the 'recipe' for mud is as varied as the recipes for chocolate brownies: a mixture of inorganic mineral particles (derived by erosion from rocks, such as quartz, feldspar, mica, gneiss) and organic material (diatoms, bacteria, microscopic meiofauna such as worms and Protozoa), with an added amorphous icing of adsorbed proteins and other organic molecules. But why are the Solway's mudflats so important in the competition between the land and the sea, both in terms of their slippery geomorphology and topography, and the mass of life that they are able to support? I realised I still needed to be shown how to look and what to look for – and to learn what those observations meant. Alison Critchlow had recently moved to Bowness-on-Solway and, as an artist who is happy working outside, was revelling in experiencing and interpreting the moods of her new environment.[1] I wanted to see it through her eyes.

We wander along the shore between Bowness and Port Carlisle at low tide, on the narrow strip that is neither marsh nor mud, but a muddle of grey pebbles and shells and the intertidal flotsam of orange-brown kelp fronds and pale flat fingers of hornwrack. Alison has been looking out at the water far beyond the wooden posts of the vanished steamer pier. She sits down at the edge of the marsh and brings out paints, water pots and her hardback notebook from her bag, and is immediately engaged in transforming what she sees into an image on the paper. 'The thing I'm loving is the gaps between the posts,' she says. 'And the posts, they're not perfectly spaced, not perfectly vertical. The gaps make sense of the bands of colour behind them.' She explains that for her 'the bands have *direction*. You see that darker band – to me it's moving left to right. The paler one is left to right too, but it's a different pace. And the greener band where the water is, it's moving right to left, a different pace again, it's coming in slower. To me that's like a different timescale – watching it beyond the posts, watching it move . . . I get it down in brief lines, it's a lovely thing to draw. It's about the timescale and colours – you see that the grey band has turned a bit green?' She has been using a broad brush to apply a dark green wash to the paper – 'this wash changes my perception entirely' – and now she squeezes the bristles to a point to add small, fine details. 'I'm not precise at all in the sketchbooks – though I might draw on top with pastels when it's dry.'

On the light: she prefers it when it's not bright and sunny. 'I love this greenish-grey light. It's unique. I'm fascinated by the light and the luminosity. It feels Scottish, it doesn't feel very English.' And, pointing, 'The wet sand, reflecting light, it's like literally stepping *into* the colour. That's quite rare.'

I ask her what colours she sees when she looks across the Firth, and as she explains I, too, begin to see them. My perception changes. 'I think of colours in terms of how I'd mix them. This closer darker, drier lumpy bit [where the pebbles are] – is purpley-grey. The mud is a mix of umber and sienna. And there's a pale blue line beyond, it's light, it's the brightest bit, brighter than the sky, surprisingly. The sandbank is pinky purple-ish; then you've got a darker, drier purpley bit again; and a slightly yellow-green stripe, where perhaps the water is deeper, it's less ruffled. And if you half-close your eyes – do you see there's a bright yellowish band above the horizon?' This takes some practice but eventually I do. 'And the distance is hazing out into the sky – it makes it hard to define the distant view.' She laughs. 'I've never really tried to talk someone through what I see. And everyone has their own ideas,' she says. 'People see colours differently. If you sit twenty people to paint something, you'll get twenty different paintings of the same place. I love that, it's endlessly fascinating.'

> . . . Forms
> Lose their function, names soak off the labels,
> And upside-down is rightways, while the eye,
> Playing at poet with a box of colours,
> Daubs its pleasures across the sky.[2]

There was a summer evening, shortly after we had moved to Cumbria, that my daughter and I were walking on the shore as the sun was setting. We stood together and watched the changing colours, and how the blood-red sun darkened and distorted as it sank behind Criffel's western shoulder. The warm hues pooled along the horizon and flared across the sky, where lemon-yellow and the soft orange of apricots reached upwards into silver and turquoise. We watched silently, bathed in the stillness and the gentle breaths of waves on the resting shore. But all the time words were flowing through my head, each jostling to be the most apt description of what we were seeing. 'Is it possible to look at that without thinking about it – can you just experience the beauty with an empty mind?' I asked my daughter. We tried: but the silent mental discourse was always there, impossible to still. I wanted to be enfolded by the moment – but 'enfolded', 'wallow' . . . the words themselves intruded.

I ask Alison about this, what happens in her mind when, for example, she watches a Solway sunset. 'I see the colours and the textures,' she says, 'I'm thinking about them as I look.' For writers, it is the words for descriptions, to pass on to others; for artists, the images of colours in the paintbox, to *reveal*.

We move further along to the edge of the mudflats by the port and now Alison sketches with a small block of polystyrene – dips it in black paint, uses an edge for lines, the end for a stippled effect: quick black squiggles and dots. 'I like the idea of using a process that's slightly random, so that it almost replicates the way the landscape forms.' There are deep runnels in the mud, their jagged edges built up layer upon layer like a three-dimensional model constructed from thick sheets of overlapping cardboard. They have a tired and ancient air, and Alison thinks they look like the folds in an elephant's legs. The mud's surface is sea-smoothed and fine, and its colour changes with the light.

*

'Any casual observer might observe the colour of intertidal sediments to change during daytime low tide.'[3] Scientific papers and reviews don't have to be written in a clinical and dispassionate style. This paper, 'The ups and downs of life in a benthic biofilm', has sub-sections like 'Algae got rhythm?', and comes from the lab at St Andrews University run by Professor David Paterson, the Sediment Ecology Research Group, more realistically nicknamed the Mud Lab.[4] A photo of the surface of mud on the banks of the (Scottish, not the English) River Eden, shows the word 'MIGRATION' spelt out in golden-brown letters against a grey-brown background. Light-sensitive single-celled organisms, diatoms, have migrated from the shaded area to cluster beneath the cut-out letters of a large stencil.

I visit the Mud Lab at the Scottish Ocean Institute, formerly and more famously known as the Gatty Marine Lab, in the summer of 2019, when it is nearing the end of extensive alterations and expansion. Between the path at the top of the beach and the building there are metal barriers and red-and-white tape, and cheerful builders in high-vis and hard hats keep giving me helpful but inaccurate directions to the well-hidden entrance. David Paterson is relaxed and welcoming, and gives me a tour of the labs – they have been 'making do' with the reduced space for many months, and boxes and equipment are piled in corners. He told me that he got into his favourite, and lifetime, topic of research when he was doing his PhD at Bath University, followed by a postdoc position at Bristol. He noticed that the mud along the Severn estuary was pigmented and patterned; he took samples for microscopy, and found layers of various diatoms causing the colouration. And this of course led to questions such as how did the layers affect the behaviour of the sediment? How could you measure the effects? And how does this relate to the stability of mudflats and saltmarshes around the coast?

Another paper from Paterson's Mud Lab suggests that the sediment-coated surface of a saltmarsh may look barren, but 'by scuffing the sole of a shoe across the surface' a green streak, a 'vivid underlying blue-green layer of cyanobacterial cells' may be revealed.[5] The surfaces of saltmarshes and mudflats are home to a thin layer of the microphytobenthos (small+plant+surface-living) or MPB. David shows me photographs and videos – you need a microscope to see the microorganisms properly. They are highly adapted for life in places where sediment is deposited, and are mainly single-celled life forms, and mainly photosynthetic like plants – using carbon dioxide from the air and sunlight to generate energy, and oxygen that helps them make carbohydrates and other macromolecules. The most dominant forms in the MPB are diatoms, or Bacillariophyceae; they are microalgae, each with a hard siliceous shell or frustule, its shape and often-elaborate decorations characteristic of its species. Different diatom species live in different environments – which is why they 'archive' the changes that have taken place in peat and riverbanks (Chapter 5). There are also the cyanobacteria, the blue-green algae that are revealed by sole-scuffing. They all know their place, because their lives depend on it.

Seen from a gyroplane, on a day when the tide is spectacularly low, the bed of the Firth seems to be exposed in all directions. Light is pouring down around us. Pale sandbanks are dappled with sunlight. Sand waves are zebra-striped with shadows and glitter, and mudflats mirror our flight with bands of silver, palest blue and violet. The light picks out cattle on the saltmarsh, creeks glimmer. But at this height we cannot see the changes that light induces beneath the surface.

Zoom in close again, down through several degrees of magnitude, to microscopic then molecular level. Saltmarshes are dynamic systems, they accrete and erode and grow upwards. They are also rather more complicated than we might suspect at first glance, because not only do the overlying plants – whether samphire or thrift or asters – know the height at which they can survive above the tideline, but so too do the microorganisms within the thin mat formed by the MPB. But for the MPB the limiting factors are not only wetness and salinity, but smothering too, for the surface of a saltmarsh is frequently inundated with new deposits of very fine, cohesive particles. This means that diffusion of oxygen, carbon dioxide and nutrients is limited – anaerobic, oxygen-free conditions can be found just a few millimetres below the surface. Nor can light penetrate far, and these microorganisms must move to find the level that suits them. Sometimes the 'classic' coloured picture of zonation can be seen by breaking open the surface. At the top are the photosynthetic golden-brown diatoms, which migrate towards the light; beneath them are the blue-green cyanobacteria which are also photosynthetic but can absorb light at lower wavelengths; and then there is a smelly layer of sulphate-reducing bacteria, which produce the hydrogen sulphide that forms a complex with iron to give the characteristic black colour and sulphurous odour of the anaerobic zone.

As for the mudflats: several years ago I joined a field trip with the Carlisle Natural History Society to visit Border Marsh. It was raining, we all had our hoods up and heads down and there was little conversation. Occasionally everyone would stop and raise their binoculars or set up their telescopes and tripods; seen from behind they formed a semi-circle of green-clad shapes with jutting elbows, some upright, some rather more stooped. The leader was carrying a spade and when we stepped out onto the mud he dug it in and raised it. There on the blade was a compact block of evil-smelling black mud, but it was riddled with the burrows of *Corophium*, some sliced in a perfect longitudinal section of the U. Mudshrimps crawled and waved their long antennae within the tubes. And around their burrows the mud was coloured a bright, ochreous brown – by irrigating their burrows the mudshrimps had oxygenated the surrounding black anoxic layer.

Looking back at photos from that day I see now what I didn't even know to look for at the time – there are patterns and patches of colour on the mudflat's surface which, if I had scraped them up and examined them microscopically, would surely have been mats of the micro-phytobenthos.

Light flooding the intertidal mudflats changes their colours, in the perceptions of both artists and scientists, and through the biochemistry of microorganisms. The wind ruffles and stirs the thin surface film of water left by the ebbing tide, and down in the mud where invertebrates stir the stagnant water in their burrows to reoxygenate it, the oxygen changes their own colours too.

A little lower on the shore than the *Corophium* burrows at Border Marsh were the coiled excretions of *Arenicola*, the lugworm, and here – as often on other shores where they burrow in muddy sand – the coils of digested sediment were black. I wonder how they can live in such foul conditions, fish-bait growing fat and red, but it is that red-bloodedness that helps them to harness oxygen from the water in their burrows. The red pigment haemoglobin circulates in their blood, not within blood cells as in vertebrate animals, but free in the fluid. In annelid worms such as earthworms and lugworms, this haemoglobin is a very large molecule and takes up oxygen from much lower concentrations than mammalian haemoglobin: it has a higher 'affinity' for oxygen. As the blood circulates through the lugworm's gills and skin, so the haemoglobin plucks dissolved oxygen out of the surrounding water. Arthropods like insects don't need oxygen-binding blood proteins because there is so much more oxygen in air than is dissolved in water – and they also have a network of tracheae, air tubes, that ramify amongst their tissues, taking gaseous oxygen deep within the body. Only aquatic insect larvae like those of some midges (often known as 'bloodworms' because of their colour) have haemoglobin. For Crustacea though, of which the majority live in water or in burrows, open air tubes are not a design option, and so the problem returns to the availability of dissolved oxygen. Crustacea are 'blue-blooded': they have a large oxygen-binding protein, haemocyanin, free in their blood. Unlike haemoglobin, which contains iron, haemo-cyanin contains copper – and when the molecule binds oxygen, it turns blue. Haemocyanin is especially important for those Crustacea like *Corophium* and the Norway lobster *Nephrops* that live in burrows. (But the pink colour that emerges on cooking is entirely unrelated; it is due to the release of astaxanthin, a carotenoid pigment normally hidden in a complex of other molecules in the skeleton.)

My friend and former colleague Professor Jim Atkinson has always been an enthusiast for scampi – *Nephrops norvegicus*, the Norway lobster, langoustine, or even more confusingly the Dublin Bay prawn (it is not a prawn). It is doubtful whether many people wonder about the animal whose tail they eat – which is not always unadulterated *Nephrops*, but can be a re-formed mixture of scampi and fish, and was at one time even monkfish tail – but *Nephrops* is a decapod crustacean, like shrimps and crabs, and its home is a burrow in the muddy seabed. Jim's lab at the former University Marine Biology Station Millport (UMBSM, pronounced 'um-buh-sum', but now – through various political and financial contortions – a Centre for the Field Studies Council), on the island of Great Cumbrae in the Firth of Clyde, was cluttered with strange Dali-esque shapes at least a half-metre high. The groups of brownish-yellow interconnected, contorted tubes seemed to be made of solidified jelly through which the light shone faintly. Each single group was a resin cast of the sub-sea home of a *Nephrops*. Jim told me that the resin had to be mixed in the boat and then passed in a plastic watering can to the diver, who then had about thirty minutes to complete the casting – by pouring the liquid resin into the burrow – before the resin set. 'Getting the casts out is often hard physical effort,' he said. 'In soft muds they can be pulled out with careful hand excava-tion to prevent breakages. At this stage you're working in zero visibility so it's best done when a slight current is running. Deep casts – some of mine are over a metre deep – require a lot of effort. I've used an air lift on occasion, even a hose from the research vessel. All good fun!' Glasgow University's Hunterian Museum is now the recipient of hundreds of his resin casts of *Nephrops* and other burrowing Crustacea.

Rather than looking at burrows *in situ* like Jim Atkinson, some researchers have chosen the warmer, drier option of persuading small species like the mudshrimp *Corophium* to burrow in experimental situations in the lab, and have then made resin casts. Others have captured the burrowing on video for later analysis. Dr Rachel Hale videoed *Corophium* gathering up 'armfuls' of sediment and moving it around, often patting the particles against its burrow walls. A layer of fluorescent pink or green particles added to the surface of the mud painted a colourful picture of how the mudshrimps moved the sediment around. But Hale and colleagues also went for a very high-tech method, computerised tomography or CT scanning (normally used in hospitals to scan brains and bodies), to analyse the shapes of the burrows of the ragworm *Hediste*, the mudsnail *Hydrobia*, and *Corophium* – as separate species, and as a mix. They set up cylindrical burrowing chambers in the lab, and then the CT scan worked its way down each cylinder, scanning 'slices' and stitching them together to make a 3D image.[6] The *Hediste* burrows branch and loop and wriggle; *Hydrobia,* being fat little snails, have very simple and short burrows; and the U-tubes of individual *Corophium* do not connect. Now that I've seen the images, I am very conscious of this hidden, complex, crowded mix (and I silently apologise for my crushing footsteps across the mud) – yet it's a mix that isn't static, because all three species are able to move around and interact.

Scuba-diving around the Cumbrae coast, another friend at UMBSM, Professor Geoff Moore, found a corophiid, *Crassicorophium bonnellii*, which does not make burrows but lives on the surface of kelp holdfasts (the seaweed's anchor-plates). It constructs tubes by sticking together small pieces of shell and other débris. Subsequently, Geoff and one of his doctoral students showed that these amphipod crustaceans produced a silk, a marvel of marine engineering – strong, flexible, long-lasting and sticky – that acts as the glue.[7] Does *Corophium volutator* produce a silk too? Geoff is confident that it does, and that the mudshrimp might be using it to stabilise its burrow walls within the mud. I like the idea of 'Solway Silk'.

New technologies can find unusual uses, whether CT scanning of burrows, or GPS to measure the changing height of saltmarshes, but some of the fun of scientific research is in 'repurposing' bits of old kit or devising the new. An important tool used by the Mud Lab researchers is the CSM, the Cohesive Strength Meter, now small and portable and commercially produced. David Paterson designed the prototype many years ago – he shows me a photo of the original version, which is tall and Heath Robinson-ish (although there is no sealing-wax or coat-hanger in sight) and built of clamp-stands and the glass cells that were normally used for a spectrometer, which is a very different piece of equipment. The CSM is used to measure the cohesiveness, and thus the strength, of the sediment's surface: it measures the pressure of water at which the surface of saltmarsh or mudflat is sheared off and eroded. This is not some merely esoteric investigation dreamed up by scientists who enjoy spending their field trips up to their knees and elbows in mud, glorious mud. It provides a crucial measure of the vulnerability of the estuarine margins to erosion. It also helps to explain the changing colours.

Diatoms and cyanobacteria in the MPB move upwards in daylight when the tide is out, in a distinct order of species, and they alter their relative positions according to changes in light and nutrients and the presence of grazers such as the snail *Hydrobia*. They also secrete a mixture of sticky mucopolysaccharides, the Extracellular Polymeric Substance, EPS, which increases the strength and stabilises the surface of the sediment; if the EPS is intact, the shear stress (measured with the CSM) required to suspend the sediment is much higher. In other words, the surface of the mudflat is more stable. But if the surface has been damaged, the microorganisms of the MPB are lifted off into the water column, and their characteristic pigments – yellow fucoxanthin, green chlorophyll c – can be detected by sensitive instruments (though not by an artist's eye).

Bioturbation and bioengineering are two useful words that relate to the activities of all these animals and microorganisms within the mud and marsh, and these words, these phenomena, affect the margins of the Firth. When a burrow is irrigated and its water aerated, it becomes surrounded by a halo of pale, oxygenated sediment, which makes the mud more habitable for other organisms that need oxygen. Also, the organic and inorganic materials within the sediment have been mixed around – think of *Corophium* shovelling and pushing particles within its burrow: the sediment has been perturbed, or undergone bioturbation. New food, new organic molecules are made available for neighbours. The bioengineers are the living organisms, large and small, who change the topography of where they live, for example by making it smoother, or changing its stability when the waves come washing in. We are learning more and more about them because not only are they fascinating and important, but there is also research funding available, due to the anthropocentric and econocentric importance of our edgelands.

One exotic and unhelpful bioengineer is the Chinese mitten crab, *Eriocheir sinensis*. I remember examining a preserved specimen when I was an undergraduate studying invertebrates; it looked harmless and slightly foolish with its 'muffs' of bristly spines – little did I realise then that the species would invade and start migrating around the British coast. It is now listed as amongst the hundred most damaging INNS.[8] It is thought to have arrived in the Thames estuary in 1935, perhaps in ballast water, and has since spread up the estuaries of the east coast; specimens have also been found in South Wales and on the south coast of Ireland (but the specimen listed for the Firth of Clyde was apparently found near a Chinese supermarket in Glasgow). It has not yet been recorded as entering the Solway Firth from the Irish Sea, but whether that is because it is absent or because no one has looked is hard to say. *Eriocheir* is catadromous – the opposite of salmon – laying its eggs offshore in salt water, the juveniles then migrating into estuaries and eventually upstream into the rivers. The bioengineering it carries out is decidedly negative with regard to sediment stability, for its vigorous burrowing undermines the edges of estuaries and river banks. Using the fluorescent particles technique, Dr Andy Blight, the lab manager of the Mud Lab at St Andrews, showed how the crab rapidly redistributed the markers within the substrate around its burrow, bioturbating and 'un-engineering' the stable sediment.[9]

A burrowing mudshrimp reaches out with its long antennae to capture detritus which it pulls in and uses to line its burrow wall. Those inorganic particles are also coated with a thin film of bacteria, diatoms and EPS, a source of food, so when it feeds, *Corophium* grabs an armload of particles between its antennae, and hauls it towards its mouth. Now its various limbs act out a highly coordinated dance, as rapid as a magician shuffling cards, with the result that particles are sifted through a basket of bristles, scraped, and the scrapings passed to the mouth and chewed. An individual mudshrimp can consume about 4,000 diatoms per hour, grinding them so thoroughly with its mandibles that no intact diatom 'shells' can be found in its gut. This is one method of feeding, and filter-feeding is another. When the tide floods over the burrow-mouth, carrying particles and diatoms, those baskets and sieves on the feeding legs can also filter out the suspended material. Imagine the effect on the mudflats' surfaces of millions of mudshrimps, feeding as the tide comes in, filtering and capturing the microscopic bounty that swirls in the water rising around them.

Out on the mudflats and the edges of the saltmarshes, plants like samphire, seagrass and the grassy INNS *Spartina* are also capturing sediment with their fronds and roots, trapping it and protecting the edges. Further out, the lugworm, *Arenicola,* is an agent of bioturbation as it eats the muddy sand and irrigates its burrow, but the sticky mucopolysaccharides with which it coats its burrow walls bioengineer some stability. Further out still, the burrowing cockles are stirring up the mud and sand. Back in the upper intertidal reaches the little MPB organisms are protecting the mudflat surface with the film of their sticky secretions – but the mudsnail *Hydrobia*, gliding over the surface on its flat muscular foot and grazing on microalgae, counteracts that. *Corophium*'s silk-lined burrows stiffen the sediment – but by scraping up and eating the protective microorganisms, the EPS and the seeds of the plant pioneers, the mudshrimp contributes to erosion.

Left to themselves and without our interventions, there is eventually a balance, between individuals, species, the size of populations, the weather and the tides. Colours and textures change.

*

At the tip of the old viaduct embankment at Bowness-on-Solway, where it juts out into the Firth between the saltmarshes and mudflats, a strange rusting contraption is half hidden amongst the gorse and dry grasses. A square metal hopper tilts onto a rectangular grid; a small wheel, which must once have had a handle, connects with a rod; the rod is attached to a system of gears which would have jiggled the grid backwards and forwards like a riddle; and at the far end there is a shute. Beneath the grid is a pile of empty mussel shells, many of them now green with algae, a few with perforations made by predacious dog-whelks. This, and a line of rotting posts offshore, is all that remains of a former mussel farm.

Further to the west though, cockles and mussels are common and natural inhabitants of the edges of the Firth, living within or just below the intertidal zone. 'Mussels and cockles are gathered along the [Galloway] shores by poor persons, and carried weekly to the markets of Dumfries and Carlisle,' according to the 1848 *Topographical, Statistical, and Historical Gazetteer of Scotland*.[10] More recent, larger-scale commercial fisheries have been closed for years because the shellfish have not reached the size and number at which they can be sustainably collected: the 'maximum sustainable yield' allows 'one-third for man, one-third for the birds, one-third for regeneration'. I had met one of the former Cumbrian mussel collectors and he had explained how the mussels were collected; hand-gathering certainly sounded like a hard way of earning a living. The gatherer used a short-handled rake to pull the mussels from a patch towards a sieve (tineal, pronounced 'teenall') or, more commonly, a semi-circular net. The mesh size was large enough to allow undersized mussels to fall through. The net or tineal was then swilled in a pool to riddle out mud, pebbles and small mussels, and the retained mussels were tipped into a carrot sack held open at the top by a bottomless bucket. The full sacks would then be carried to a collection point.

Back in early 2005 and on a very low spring tide, Dr Jane Lancaster and I walked out onto the mussel beds of Ellison's Scaur on the English coast. At that time Jane, a marine biologist, helped the Cumbria Sea Fisheries Committee with the annual shellfish surveys.[11] Mussels are bivalve molluscs, *Mytilus edulis*, that obtain their food by filtering it out of the water using cilia, fine flexible extensions of their cells, on their broad gills. I had not previously associated mussels with mud, thinking of them as animals that need to attach to rocks and pebbles. I certainly knew nothing about 'mussel mud'. As we strode over pebbles, mussels and the fairly firm sand of the inner scaur, Jane told me that a few years previously this area had been knee-deep in mussel mud. 'It's very fine, sticky and soft, and you sink right in over your knees. In the summer this will be impossible to walk over.' The mud is a mixture of trapped silt, mussel faeces, and pseudofaeces – the filtered particles that are rejected as inedible. Astonishingly, it accumulates at the rate of about fifteen centimetres a year, and young mussels 'produce vast amounts of the stuff'. Sometimes the mussel mud is thick, but then the sea removes it; sometimes the mussels themselves are densely packed, at other times they are torn off and swept away, cast up in clumps on the upper shore. On the mussel beds we found circular patches which had been stripped bare by the sea, and in other areas there were empty shells, all sizes.

Wading birds like oystercatchers and turnstones prise open and eat the smaller mussels; further out the starfish prey on them. Jane had once counted starfishes at an average density of seventeen per square metre, and told me 'They can decimate a mussel bed in weeks'. By early evening we were out in the middle bed, crunching over the animals and splashing through pools. Black bivalves spread in all directions over the lumpy terrain; mussels in metre-high domes or lying flat aligned with the currents. There were whelks and winkles and green algae, too; waders were running and probing and twittering on every side. We watched an oystercatcher weaving away from competitors, with a mussel lodged sideways in its beak.

The lagoons would continue to drain for about an hour after the tide started coming in, and at the moment the water was pouring out over the lips of the pools, very fast and purposefully. But if the water in the channels started to have scum on its surface, we would know the tide was coming in. We were a mile or more out on the middle bed of the scaur when the tide was due to turn, but Jane appeared unconcerned and indefatigable (half an hour later I found myself suggesting, rather weakly, that 'perhaps we should go back now?', and she was kind enough to humour me). The mussels in this middle bed looked in good condition, about thirty millimetres long, their shells thin with brown stripes; they were a year old and would grow fast now that winter was over. The mussels in the outer bed would grow even faster because they are covered by water and so can feed for longer periods. So where do they all come from? Mussels release their eggs and sperm into the water at mating time and the small motile larvae swim about until they find a place to settle and attach. The larvae come not only from local mussels but – and here the Upper Solway is especially fortunate – are also recruited in huge numbers from the south, swept in from the Irish Sea by longshore drift. Larvae are small enough to be filtered out of the water and eaten by the adults, so the safest places for them to settle are bare areas on a solid substratum (for example, when Ellison's outer bed had been scoured clean by storms two years previously, the cleared space was heavily colonised by spat). While we were talking, Jane suddenly bent down and picked up a piece of red filamentous seaweed and teased apart the fronds. 'There are the spat, do you see?' And there they were – dozens of shiny pale grey shells as small as fleas, attached to the weed; the spat would over-winter amongst these red algae and in the early spring would let go and settle on the mussel beds.

This, like mussel mud, was a revelation. Mussels fix themselves to a hard surface by secreting thin brown threads, the byssus, with a sticky free end, but now I learnt that the attachment is not permanent: here were minute spat that could release themselves and be carried elsewhere to settle anew. During hand-gathering, undersized mussels are discarded and left to grow. So could they, too, reattach? 'Even while you're sieving,' Jane told me, 'some of the small ones will have reattached to the sides of the sieve!' The byssus and the 'glue' that attaches the thread to the substratum can clearly be secreted very quickly, extruded through a channel on the foot, and the byssus can be lengthened too. Because of the silty, unstable nature of the Solway beds, the mussels may be anchored as much as a metre below the surface; as mussel mud builds up, the byssal threads are extended, and the animals can also pull themselves down into the mud for protection during a storm. The byssal thread is made up of interesting proteins: those nearest the mussel are very elastic, like a soft extensible spring – when a wave tugs at the mussel, the byssus stretches. But the danger is that the elastic recoil would dash the mussel against its anchoring rock. To avoid this outcome, the end of the byssal thread close to the rock is made of stiffer proteins: an elegant solution to a tricky engineering problem.

There was another, more unusual, use for byssal threads – or 'sea silk' as they are known. Byssus from the large, slow-growing Mediterranean clam, *Pinna nobilis*, used to be collected, carded and woven into expensive garments. Marine biologist Helen Scales tells, in her book about seashells, *Spirals in Time,* how she went to Sardinia to find out more about this dying craft and was shown untreated byssus, 'a knotty tangle of threads embedded with tiny seashells and blades of seagrass, like the ginger beard of an old man of the sea, flecked with dinner', followed by a 'tuft of soft golden fibres that gleam in the sunshine. This is clean and carded byssus, ready to be spun. This is sea-silk.'[12] A small turban-style hat was sold recently at an auction in New York; it is estimated that the byssus of eighty clams might have been used to make it – so it is fortunate that *Pinna* is now internationally protected.[13]

On the Solway, *Mytilus* are very occasionally found to contain small, misshapen pearls but, as a former mussel gatherer told me, 'Don't waste your time hoping to make your fortune from pearls on the Solway shore – you'd need more than a hundred to cover a penny!' (In contrast,

I and my family have a little matchbox of pearls we collected from mussels somewhere on a grittier Scottish north-west coast – small, misshapen and odd colours, but a beautiful example of how mussel cells take calcium and carbonate ions out of the water and convert them into nacreous crystals.)

The mussel beds on the English side, and on the Scottish side – on The Rack that leads to Heston Island in Auchencairn Bay, and around Rough Island near Kippford – are now all closed. The last survey at Ellison's Scaur and along the coast towards Silloth, carried out by the North Western Inshore Fisheries and Conservation Association in May 2018, found insufficient numbers to reopen the fishery. The reasons for the decline are unclear.

Common cockles (*Cerastoderma edulis*), like mussels, are bivalve molluscs that filter-feed, but unlike mussels they live in shallow burrows actually within mud and muddy sand. The cockle fisheries on both sides of the Solway are closed too. On the English side, there has been no commercial cockling in the Mawbray and Silloth area for at least fifteen years and it has been at least ten years since cockling was permitted at Auchencairn; when Marine Scotland last surveyed the site in 2010 the stocks were still unsustainable for commercial harvesting. And yet, if you drive along the edge of beautiful Auchencairn Bay you are suddenly surprised by bright, white patches close to the road near Balcary, as though sunbeams had lit up a section of shore. But it isn't sand – within a few steps from the road, you are crunching through drifts of empty cockle shells. The cockle drifts are even more extensive and dramatic at Kippford Bay further along the coast to the east.

There was an aeolian orchestra at Kippford. Rumbles of bassoons, tinny clankings and a prolonged hissing like a pump that has sprung a leak – the wind blowing up the Urr estuary was playing the fittings of the boats that were standing on the hard. The tide was out and so there were mudflats, overlooked by the row of sturdy houses and cottages. The colours of Kippford are different from the sandstone country, here they are the pale greys and pinkish greys of granite. Granite makes the houses, the rockeries, the steep lush gardens and the drystone walls that are a mixture of sharp edges and water-rounded boulders, decorated with yellow lichen. There is a strange little monument of rough stones and an engraved polished slab, supporting a tall, oddly shaped piece of metal: 'Relic of the Kippford granite industry'. Then, around the corner, the muddy margin of the estuary has suddenly been replaced by shoulders and platforms of pink granite that emerge like islands from a buttery, creamy-white sea of shells. Cockles! John and I walk towards the bay, the wind and showery rain in our faces, and the cockle drifts get wider, deeper. The ridged and globular shells are mainly from animals that were full-grown, but there is a sprinkling of medium-sized and small shells too. I scuff my boot across the surface then dig down, and there is a deeper layer of fragments, some as fine as sand, where the shells have been abraded and crunched by the tides. The bank of shells rises more than three metres high; it has engulfed the roots of the stunted hawthorns at the top of the shore, and encircled their trunks. Millions, billions of empty shells.

How long have the shells been accumulating? Why have so many cockles died? I have asked so many people, but there are no real answers. 'They have always been there', 'There have been masses of cockle-shells for many years', 'They've increased in the last ten years' . . . The cockle fishery was closed eleven years ago, so 'perhaps the increase in numbers shows the stocks are recovering?' A survey carried out in the Glenisle and Rough Island area in 2015 found that about 50 per cent of the live cockles were small and young – presumably the older ones are piled up on Kippford beach.[14]

Out on the intertidal plain of shelly, muddy sand that extends towards Rough Island I scrape the surface and dig with my heel (the trowel had been forgotten), and see that the black anaerobic layer is only a few centimetres below the surface. Although I cannot see them, here and at Auchencairn Bay the flats will be perforated with millions of living, growing, dying cockles. Each was brought here by the tide, reaching the estuaries' edges as a minute post-larva, with a foot and tiny transparent shells less than a millimetre long. Each settled on the mud and began to burrow down to safety.

Philip Henry Gosse is so very quotable. His friend the Rev. Charles Kingsley, the minister and writer who was also an enthusiast for marine creatures, had dredged some cockles in Torbay and sent them in a hamper to Gosse in London, his letter requesting that their 'respiration should be worked out'. These large, spiny cockles, *Cardium tuberculatum*, disturbed the Gosse family's peaceful evening.

Many persons are aware that the Common Cockle can perform gymnastic feats of no mean celebrity, but the evolutions of Signor Tuberculato are worth seeing. Some of the troupe I had put into a pan of seawater . . . [and] by and by, as we were quietly reading, our attention was attracted to the table where the dish was placed, by a rattling uproar, as if flint stones were rolling one over the other about the dish. We could look at nothing but the magnificent foot, and the curious manner in which it was used . . . [It] is suddenly thrust out sideways . . . then, its point being curved backwards, the animal pushes strongly against any opposing object, by the resistance of which the whole animal, shell and all, makes a considerable step forwards. Cooped up with its fellows in a deep dish, all these herculean efforts availed only to knock the massive shells against the sides, or to roll them irregularly over each other. A considerable number of those sent up we 'killed to save their lives'; making gastronomical use of them . . .[15]

A cockle burrows by thrusting its muscular foot out between the valves of its shell, then expanding the foot like an anchor, and drawing itself down. As it burrows it rocks its shells to widen the burrow, meanwhile keeping an opening to the surface from which, when the animal is not quiescent, its tubular siphon projects. Water is sucked down the siphon and circulated over the gills; oxygen is extracted, and food particles (plankton, diatoms) suspended in the water are trapped and sorted. The burrows of even the older common cockles are only a few centimetres deep, and the cockles move house frequently, so they are often vulnerable. Flounders, and waders like knot, prey on the small cockles near the surface, oystercatchers and shore crabs prey on the larger ones. The globular shells can be disturbed and rolled by the waves, and cockles that have been exposed on the surface are very vulnerable to extremes of temperature, especially in wide, shallow bays. But no mass deaths have been reported; there are no piles of stinking corpses. I've missed the best low tides, and the short days are here, so it's too late this year to get to Rough Island or Heston Island. But next year I will walk out there with someone who knows about cockles and mud.

*

It's a low, low tide at Moricambe Bay. I don't think I have ever seen so much of the Bay's bed revealed, and the edges of nearby Skinburness Marsh are steep dark cliffs at the edge of the dazzling flatness. Change the scale and these could be towering cliffs at the edge of an ocean. There is so much sky: this is such a feature of the Upper Solway, but today it seems that three-quarters of our view is made of clouds and blue light, clouds of all types and shapes and densities that range from dazzling white to silky grey. Lionel Playford has walked out along Grune Point with me. 'It's my kind of day!' he says. Lionel is a land-scape artist who likes to paint outdoors, often using natural materials from his environment, and one of his recent projects, *Sky Gathering*, was with the Cloud Appreciation Society on Lundy Island.[16] Lionel and I first met a few years ago during a conference about peat; we were chatting as we walked across the raised bog of Bowness Common and we had both, independently, noticed that two twigs on a nearby hawthorn tree were shaking. No bird had flown up, and it was the wind that had set them vibrating, but at different frequencies. Almost simultaneously we each commented on the twigs 'having different resonances'. That sounds ridiculously abstruse now, though it made us laugh at the time – but the remarks came out of our backgrounds in science and in Lionel's case, engineering (he was formerly a naval architect, designing and building ships at Barrow). Today we are at Grune Point to look at mud, because I want to understand how Lionel the artist 'sees' mud, and also to talk a little about how scientists and artists can work together. Years ago I had been very involved with the ideas and practice of 'SciArt' – working with a sculptor, the late Rebecca Nassauer,[17] and running national projects and conferences that brought scientists, artists and fiction writers together; Lionel, as part of his PhD on art and climate science, had been resident artist, teacher and researcher aboard the German research icebreaker *Polarstern* as the scientists were gathering evidence about climate breakdown.

So we walk along the edge of the marsh, past Grune Gutter and Calvo Creek where fresh water trickles seawards at the bottom of deep channels, and we talk about what we see. There is an island of sandy mud that is shaped like a perfectly symmetrical teardrop or, since its sides are patterned with ripples, a fish with a long tail; the tail is pointing upstream relative to the beck but downstream relative to the way the incoming tide would advance. Which force of water has the greater influence? But their influences will change day by day, perhaps hour by hour. We wait and watch, and it is a long time before the incoming tide enters the Bay, but suddenly it is there, everywhere. The tide pushes upstream around the island; there are short poppling waves hitting its seaward edge and the tide is embracing it on both sides and forming interference patterns where both arms meet. Water is swirling onto the island from front and rear. A small bay at the foot of the mudflat where we are standing fills with water and there is a spiral of foam gyrating on the surface; the water there is flat in contrast to the nearby rippling, and when the foam spiral disperses the flatness remains. The water that has covered the teardrop island is a smooth plate amongst the roughness. The leading edge of the tide must be picking up microorganisms, living and dead, and the organic molecules that they have made like proteins and fatty acids, and these materials are being gently churned into a froth. The yellowish foam is accumulating rapidly along the tide's edge and we speculate about the foam's effect on surface tension, like 'casting oil upon the water'.

Lionel sits on the edge of the saltmarsh and draws. He wets the paper, and smears bands across it, using mud from the bottom of the shore – 'lovely slimy stuff' – and drier mud from higher up. He spreads the mud with a knife blade and then uses a blue draughtsman's crayon to draw over it. 'See how the colour on the paper isn't the colour of the mud on the shore,' he says. 'It becomes yellower, more ochre-y.' Earlier, boot stuck in the mud, I had fallen on my hands and knees and the mud on my waterproof trousers is drying to that same pale yellow-grey; but it is also, intriguingly, sparkling with minute flakes of mica.

Lionel's impression of what he sees is rapidly emerging on the paper and he has perfectly captured the movement. He tells me that his method is to use a 'first pass', a glance at the whole, and then to concentrate on the detail. 'When I look at an event that is changing, I obviously cannot deal with it all so I have to choose what to focus on. Then follow that through and study the pattern. I need to *suggest* what is happening, so that if I look back on the sketch days, weeks, months later, it still brings back the dynamic.' He points out the binary colour of the waves, brown on the near-side where the light is shining through and diffusing in the suspended sediment, but on their further sides reflecting the sky, a flickering pale blue. It's good to be made to think, 'Of course!'

Now there is a gentle splashing of small waves against the shore where previously was bubbling white noise, and a new surge of tide is coming in, dark-edged and with bigger ripples; it looks as though it is higher and riding above the rest. We talk about currents and friction, and the Bernoulli effect (leading to the phenomenon of 'squat' in moving ships[18]) and argue about environmental writer Barry Lopez's assertion that 'Whenever we seek to take swift and efficient possession of places new to us, places we neither own nor understand, our first and often only assessment is a scientific one.'[19]

Our discussion is hardly worthy of the Royal Society, because there are continual distractions: two young friends of Lionel's family are chatting and drawing nearby, his charming but badly behaved dog frequently sprays us with mud and water, and a strangely dark-clothed man with a balaclava has disappeared into the bushes behind us on the Point (when I encounter him later he points out two pintail ducks that

I had missed, so he is probably a birder). And, after I have arisen from my fully clothed dive into the mud, we also talk about landscape artist Andy Goldsworthy's early performance art where he wallowed naked in the mud of the Lancashire coast.

As Lionel starts a new sketch, I notice that the mud in front of us is prickled with coruscating light. We had seen that the mudflat was densely packed with the burrows and entrance-cones of *Corophium* – and now the mudshrimps are reaching out and waving! All across the mud, they are waving their antennae – they are not scraping up food, but seem to be signalling or sensing the air. Each is rapidly extending and then withdrawing its long antennae, and the incident sunlight is catching and highlighting the movement. It is another 'Lopez moment': suddenly, 'You know that the land knows you are there.'

8

Seafood

There was a low spring tide, down to less than a metre above Chart Datum,[1] so I had taken a group of people to the Allonby Bay Marine Conservation Zone[2] to wander amongst the rocky scaurs and the honeycomb-worm reefs on the lower shore. We were back in the car park and I was handing round photos of the various polychaete worms that were hidden inside the tubes and burrows that we had seen – the honeycomb worm *Sabellaria*, the mason worm *Lanice*, the lugworm *Arenicola*, the trumpet worm *Pectinaria*. We were admiring the colours of their tentacles and gills, talking about how they made their tubes, and thinking about where they lived. I mentioned the musicality of the scientific names of so many marine polychaetes: *Harmothöe, Ophelia, Eunice, Aphrodite* (Crustacea have them too: *Eurydice pulchra* – Janet Baker's lament from Gluck's *Orfeo* always playing in my mind).

And then someone asked, 'But what's the *point* of them?' The question was startlingly simple. But how do you begin to answer a question like this, that derives from a completely anthropocentric view of the surrounding world, in this case of an intertidal shore where the web of interactions between living creatures is so visible? There are so many possible answers, scientific, artistic, ethical and philosophical . . . But, more prosaically, someone in the group mentioned food chains: the worms are there as food for other species – for fish and crabs and birds. In other words, worms exist to be fodder.

MARLin, the online Marine Life Information Network, is a valuable source of information about the biology of marine organisms – not least about which species eat each other. For example, if the lugworm *Arenicola* is so careless as to protrude the rear end of its body from its burrow while it's defaecating, it stands a good chance of a passing ragworm or a plaice nipping off a few segments. As for mudshrimps, MARLin lists their predators in detail (I've omitted the references to scientific papers, but you can find them online): '*Corophium volutator* is an important food source for dunlin (*Calidris alpina*), redshank (*Tringa totanus*), shelduck (*Tadorna tadorna*) and flounder (*Platichthys flesus*) and these predators can consume 55% of annual *Corophium volutator* production. *Corophium volutator* is also fed upon by the brown shrimp (*Crangon crangon*) and the green shore crab (*Carcinus maenas*), which can consume 57% and 19% of *Corophium volutator* production respectively . . . *Corophium volutator* has the habit of swimming when immersed, which makes them available as prey for the common goby (*Pomatoschistus microps*), herring (*Clupea harengus*), sprat (*Sprattus sprattus*) and smelt (*Osmerus eperlanus*).'[3]

Those predators: the birds – dunlin and redshank; the fish – flounder, herring and smelt (sparling); the invertebrates – the brown shrimp and shore crabs: they are all part of the stories of the mudshrimps and the Solway.

*

Ronnie Porter, who once took me along the Allonby shore to point out the named scaurs and boulders (more on this later), lives in an immaculately cared for terraced house that was once part of a herring shed. We meet again to talk about fishing, and he brings out a red hardbacked file bulging with plastic folders that contain old postcards and photographs of the area. There are a couple of pictures of a long building with a blank un-windowed wall and no doors, and just one entrance at the end. Now the building is a row of houses with a shared courtyard out at the back. A tall three-storey building at the northern end was the Solway Hotel, which his aunt owned and which was burnt down (twice). The Porters' house was part of the fish yard, where the herring was salted and stored.

'Allonby people in the old days,' he tells me, referring to when he was younger, 'they were farmers, smallholders and fishermen. Herring was the main thing, they'd spawn and shoal in Allonby Bay. Most people in the village had nets.'

The nets were about six feet high and as much as twenty yards long, and were set down on the sands at low water when there was a 'big tide' ('big tide' for local people refers to the lowness rather than the height of the tide).

'They had cork floats along the top and lead along the bottom. Then there was a big bunch of lead and corks at one end. At the other, more corks and a pole which floated upright. There was a rope til an anchor, just at one end – so the net would swing around it with the tide. One-inch mesh – gill nets.'

Everybody had nets, some people had two or three; Ronnie and his friends had their own 'bits o' nets' when they were young in the 1950s. The nets, which were heavy even when empty and even heavier when full of fish and sand and weed, were taken down the shore and later collected in hand carts or by horse and cart.

'It was mainly a farming community then so horses were around anyway. Grandad was a farmer, he had a horse. There was an old lady up at White Lodge – she was a hard old devil, she'd pick up and put down the nets herself! There wasn't much time to empty the nets – you'd bring them up full, empty them out on the green and put them back down.'

Like Neil Gunn's fishermen, who 'forgot everything, except the herrings, the lithe silver fish, the swift flashing ones, hundreds and thousands of them, the silver darlings',[4] even now, after all these years, Ronnie's eyes glimmer when he talks about the herring. 'It was an amazing sight. A sunny morning, and the sun shining on the herring – sometimes they'd still be in the shallow water, and shining. I've seen the nets chock-a-block with herring . . . I remember going down with

Grandad one April and it was blowing a gale – the nets used to drag, and that many with herring could end up on Dubmill Scaur, dragged from the other side of [the rocky scaur] Popple.' Didn't the nets get tangled with each other? 'No, they knew where to put them. Most people had their own bit of sand. Grandad was near Bank House, Twentymans further along, then Thomsons.'

Fish merchants would come up from Maryport with their wagons to take away fresh herring, but most of the catch was salted in Allonby. The salt came in 'great big blocks. I used to chop it up for Grandad with a hammer.' The fish and salt were layered into barrels for storage; drawings in the Holme St Cuthbert's History Group's delightful book of photos and other images, local history and local reminiscences, *More Plain People*, show the women at work salting and packing herring. (This book, and the first *Plain People* are probably my favourite books about the past and recent history of people living on the Cumbrian Solway Plain; there is so much precious, personal information recorded which would otherwise have been lost.)[5]

The herring mainly came up the Firth to shoal in April and September. Ronnie thinks they were there most of the time 'but it was said they went in ten-year cycles – ten years good, ten years lean. Then the big trawlers came in the Irish Sea and that was the end of them . . .' It is several decades since a shoal of herring was seen in the Solway. The mudshrimps would not have been at risk of predation from the shoals, those fiercely ecstatic murmurations of flashing fins and pointing snouts, their streamlined synchronised swooping. It would be inappropriate for a fish in that dense community to show its individuality and leave, to grub along the bottom. When they are in feeding mode, the herring move up and down in the water column throughout the day, the adults feeding mainly on plankton; it is the young herring who browse the 'nektobenthos' where the burrowing and scurrying animals roam.

In the late eighteenth century Solway fishermen were chasing herring not just in the Firth, but as far away as the Hebrides and the west coast of Scotland; the hunt was made easier in 1756 when 'large shoals of herring came into the Solway Firth'. Joseph Huddart's father, who was a farmer and shoemaker in Allonby, decided he'd try his fortune with fish and – according to Joseph's short biography in *Memoirs of the Distinguished Men of Science Living in the Years 1807–8* – started to trade 'in conjunction with a Herring Fishery Company, while his son took his place with others in the boats'.[6] This rather understates the situation. William Huddart, a direct descendant, in a meticulously researched and detailed biography of Joseph, explains that Joseph's father, William, and some 'respectable' neighbours set up a company to exploit this influx of fish, building a fish yard on the Allonby shore. When it was sold in 1778, the advertisement described a 'large and convenient FISH-YARD or FISH-CURING CONVENIENCE . . . and on another range, hewn stone cisterns that would hold at first pickle 500 barrels of herring; with salt houses and other offices suiting the premises. Also 5 Smoak Houses for drying Red Herring, on a good situation that will hang at one time 500 barrels . . .'[7]

Joseph was only in his mid-teens when he started going out in the herring boats, but by the time he was twenty-one he was commanding the sloop *Allonby* to carry the company's cured fish to Ireland. These regular voyages made him very familiar with St George's Channel in the Irish Sea and, because current charts were inaccurate, he made a new and detailed hydrographic survey of the channel. His father died in 1762 and left him a share of the company, and just a few years later Joseph commissioned his own boat, a collier – to be named *Patience* after his aunt – from the Woods' well-established shipbuilding yard at Maryport. In this brig he took coal from the Cumberland coalfields to Ireland and North America, on one occasion returning from Massachusetts with a cargo of timber for Whitehaven.

Joseph Huddart's story is extraordinary – years of commanding ships, years with the East India Company and Trinity House, years charting the seas around the Hebrides and parts of the coast of India and China. Taught by a clergyman in the small school at Allonby, he went on to become a proficient mathematician and inventor. Sufficiently esteemed to be elected a Fellow of the Royal Society, he sits amongst the group of 'Distinguished Men of Science' in William Walker's engraving of 1862 – a group that includes Marc Kingdom Brunel (Isambard's father), James Watt, Matthew Boulton and Thomas Telford. One of Huddart's most important inventions was in cordage, developing rope which would not fray under stress; his new technique used a Boulton & Watt steam engine to twist the yarns. He lived to 'an advanced old age' (1740–1816) and when he became seriously ill at the end, he 'turned his active and comprehensive mind to the study of the anatomy of the human body, with which he soon made himself acquainted', according to William's account. It seems he was a modest and hard-working man, and much liked. The *Times* obituary describes Captain Huddart as 'tall and erect, his features were regular, and his countenance strongly indicative of those powers of mind for patient investigation and rational conclusion . . . blended with an expression of placid benevolence equally characteristic of that amiable simplicity which so strongly endeared him . . .'[8]

He lived in London after he left the sea and at his own request was interred in a vault beneath St Martins-in-the-Field in Trafalgar Square– but in Allonby's Christ Church there is a marble plaque in memory of this 'famous son' of the village. The church is kept locked so I had to ask a local lady for the key. Appropriately, she also runs the local history society and is a fan of Joseph Huddart; the society is hoping to put a blue plaque on his house – and on the house where artist Percy Kelly lived (for if the Ship Hotel can have one for Dickens' and Wilkie Collins' short stay, how much worthier are Huddart's and Kelly's homes?). The light floods into the church through the plain side windows and Joseph's plaque is impossible to miss because it is so large, and there are so many words beneath the high relief of his head. The engraved words are small and I have to stand on a pew to read them:

> . . . And of him it may truly be said that the pre-eminent powers of his mind, and his superior acquirements of Mathematics, Mechanics, and Astronomy were unceasingly devoted to the services of humanity, by pointing out a More Secure Path in the

Trackless Deep And by encreasing the facilities, and lessening the dangers of those who 'go down to the sea in Ships, and occupy their business in great waters'.

THESE MEN SEE THE WONDERS OF THE DEEP . . .

His profile is of a beak-nosed man, hatless, with curling hair that doesn't quite hide an unflatteringly large ear-hole, and piled around him are symbols of his skills – a sextant, telescope, globe, ruler, paper with geometrical drawings, scroll, an anchor, and an unidentifiable object that looks like a scimitar. It is an impressive memorial of greyish-white Italian marble, commissioned by his son and carved by 'Petrus Fontana, Carara 1821'. But despite his rise to fame in the maritime and scientific fields, Joseph Huddart is still often remembered in Allonby for his interest in herring.

*

Twelve rivers empty their fresh water into the Upper Solway Firth. The Firth's salinity is therefore always changing depending on the state of the tide and the amount of rainfall on the Lake District Fells and the Scottish Southern Uplands, and this determines the types of wildlife that live on its shores and in its waters. Salinity – the amount of salt dissolved in seawater – is described in 'parts per thousand' rather than as a percentage, and in the open sea is generally thirty-five parts per thousand. Near river outflows in an estuary it may drop down to single figures, and only animals that have evolved to regulate their own internal concentration of salts can live there. The ragworm, *Hediste diversicolor*, a common neighbour of the mudshrimp, can live in salinities as low as 4–5 parts per thousand; the mudshrimp is similarly tolerant. Fish like sparling and salmon switch their metabolism when they leave the sea to spawn in rivers, but fish like the herring are not freshwater-tolerant at all.

For humans, salt has always been a precious commodity, for taste and for preservation, for personal use and to be traded, irrespective of whether it is extracted from seawater or mined, and we have seen how important it was for Allonby's herring fishermen. On the coast just south of Allonby, and below Swarthy Hill with its remnants of Roman mile-fortlet 21, are the Crosscanonby salt pans. There is documented evidence for saltworks (salt pans, salterns) along the Cumbrian coast from the head of the Solway right down to Millom, but the Crosscanonby salt pans are the best preserved, although at continual threat of erosion – even the sea defences of stone-filled gabions are battered and broken. We flew over them during my gyroplane flight (Introduction) and the outlines of the large circular pit and lumpy waste heaps right next to the shore were clearly seen. On the ground, overgrowth by brambles obscures the details, and occasional work parties of Solway Coast AONB volunteers come armed with loppers and strimmers to expose them again. The stone walls and base of the pans are now partly hidden by grass, but by scuffing around it is possible to find burnt brick and clinker in the waste heaps, and several years ago it was still possible to see wooden stakes out on the shore, marking the position of a seawater tank and pumping system. There are some good words, like *sleech* and *kinch,* associated with salterns. In sleeching, salty sand and silt are collected from the beach into a large pond, and it is then kinched – seawater is pumped or allowed to flow through it. The resulting concentrated brine is collected in a brine pit, and boiled in iron pans to crystallise out the salt. It's not clear whether the Crosscanonby saltern used sleech, but seawater was obviously pumped up into the large circular pan, and after storage and some evaporation, drained into the now-indistinct brine pit, and taken from there to a pan-house, where it was boiled. The fuel for running the pump and heating the water was coal, for it was plentiful in this region, although an article by John Martin on the Cumbria Industrial History Society's website suggests that the ash-heap shows 'that only the poorest quality of coal must have been used as it contains such a large amount of fused material and burnt slate'.[9] The Crosscanonby salt pans are post-medieval, set up by the Senhouse family from Maryport, and were in use from at least 1684 until about 1790.

However, salterns had been used both sides of the Firth since much earlier times, hinted at by placenames like Salta, Saltcoats, Saltcote, even Southerness or Salterness. The Cistercian monks of Holme Cultram Abbey near Skinburness were major developers of the industry. They pioneered farming and fishing in the area, and of course this meat and fish needed to be preserved by salting. Living close to the margins of shallow Moricambe Bay, they were ideally placed to set up coastal salt pans and, according to John Martin, in '1536 they were farming 21 pans stretching from Saltcotes to Angerton and the Border, and everyone carried the right to cut turf [peat]'. The abbey was not far from the Mosses (Chapter 6) so peat for heating the saltwater or salty sand was plentiful. Excavations by Grampus Heritage and the West Cumbria Archaeological Society (WCAS) at the abbey are revealing more information about the salt pans, and even how the monks had rapid access to the sea (Chapter 7). Back in 2015 when I went to look at the dig, Pat Bull – a former president of WCAS – kindly showed me some of the finds. She slipped a tiny metal disc out of its polythene envelope and showed me the simple depiction of a ship on one side and the letters 'S E L' on the other; it was possibly a salt token; several of these had already been found. Another object on the table was a fine pin with a rounded head, for fixing a shroud. Pat opened packet after packet, showing me metallic fragments – a piece of leading from around stained glass, a book clasp, a tiny spur, something that looked like a mustard spoon (or perhaps it was for removing earwax), and fragments of coloured window glass and pottery. To be able to hold them and feel their weight was so different from seeing them in a museum collection. It restored them to practicality, as objects that were useful and *had been used,* rather than *objets* to be catalogued and dated, made abstract.

More recently, the archaeologists have also been turning their attention to investigating the monks' methods of salt production, which likely carried on for nearly 300 years between the abbey's foundation in 1150 and its dissolution in the 1530s. Mark Graham of Grampus told me that they had done a walk-over survey along the saltmarshes on the south side of Moricambe Bay, from Grune Point to the River Waver, and had found several circular ponds. Little is known about how the monks concentrated the salt from seawater. Mark conjectures that these circular ponds, all at the landward margins of the marshes, could be brine pits – seawater flowed up channels in the marsh and was trapped in the ponds by sluices. How the monks then treated it, and where, is not yet known, although it is known that they had turbary – peat cutting – rights at Wedholme Flow, and sandstone clapperbridges have been found across ditches on the marsh and at the Flow (Chapter 7). In fact, the Solway saltmarshes were perfect for salt production, and the Industrial History of Cumbria website notes that archaeologists have found post-medieval saltworks all around Moricambe Bay, from the Skinburness Marsh to Border Marsh to Newton Arlosh Marsh, then around the coast of the Anthorn peninsula up to Burgh.

As for saltworks on the Scottish side, Neilson's 1899 *Annals of the Solway* records that the Holme Cultram Cistercians acquired saltworks at Rainpatrick near Redkirk from their mother house, Melrose Abbey, but physical evidence no longer remains.[10] The remit of the Solway Coastwise project, run by the cross-border Solway Firth Partnership, was to look at the origins of placenames on the Scottish coast, and to investigate various archaeological features. Nic Coombey's office is an eclectic mix of maps, shells, bones (although not the mammoth bone, see Chapter 2), photographs, books on identification of marine organisms, books on Solway history, and more, and we spent a happy morning talking about shrimpers, harbours, the names of rocks and caves – and salt pans. Nic showed me his notes from a report from 1714 by the Comptroller and Collector of Customs at Dumfries that mentions salt pans at Rascarral, and two at Priestside, where there had 'formerly been a great number'.[11] Nic also pointed out where 'salt pan' was marked on an 1870 sketch-map of the Rascarrel Estate, but when he and archaeologists from St Andrews went to investigate it turned out to be a natural pool in the intertidal rocks, with an enclosing cobble wall – perhaps a 'bucket pool' where water was collected, then pumped or carried to the salt pan building. As Nic explained, this is something of an anomaly, because salt had been imported cheaply since the mid-1800s and salterns had gone out of use.

But I often wonder if oddities like this exist because someone just had the idea, 'This will be fun – let's do it!' There are records – but little remaining evidence – of other saltworks along the Dumfriesshire and Galloway coasts. At Caerlaverock there are are examples of saltcot hills, although they are more like mounds. Nic explained that once the salt was removed from the sleech, the remaining mud and sand were excavated from the pits and cast aside, creating mounds that grew into hills as years of salt making continued. I wished I'd read the *Annals* before I visited Brow Well to see where poor Burns had drunk the waters, because Neilson mentions black-bottomed pits or kinches along the top of that merse: 'there is no growth or vegetation in these lifeless holes. They are thus no unfit memorials of a dead industry.'

After Mark Graham had shown me the Moricambe Bay ponds on the Ordnance Survey map, the need to go and see them was of course irresistible. It was a blue and blustery mid-August day, already feeling autumnal, with blackberries and haws ripening in the hedges. The wind was roaring through the thickets and the skinny profiles of ash trees – ash dieback disease is already dramatically changing the outlines of the county – and was stirring waves of motion amongst the sedges and tall pale grasses of Calvo Marsh. In the distance, black-and-white Friesians were grazing the saltmarsh stints of Newton Arlosh. There were two blue circles marked on the map and the ponds were easily found, right by the road but protected by barbed-wire fences. Their edges were thickly colonised by rushes, and lumpy patches of floating weed obscured the surface, but there still remained enough water to reflect the white pillars of cumulus towering into the blue sky. A ditch ran close to one of the ponds, and those perfect circles had to be man-made, but I doubted whether the Solway's tides could still reach them. These saltmarshes on the south side of Moricambe Bay are protected by the jutting peninsula of Grune Point, and have likely grown considerably in the centuries since the abbey was dissolved. It is a puzzle which can be left for the archaeologists to solve.

*

These days, predation of *Corophium* by herring and smelt (but since they are known as sparling on the Scottish side of the Firth, I'll use that name – 'smelt' is too easily confused with 'smolt', young salmon) probably has little effect on the mudshrimp populations: herring has all but vanished, and sparling numbers are much reduced all around the country's shores.

Sparling, *Osmerus eperlanus*, anadromous fish like salmon, are tolerant of a wide range of salinity; as adults they live at sea, but they return to the fresh water of the rivers to spawn. Shoaling in the mouths of estuaries in the autumn, they leave the sea in early spring, some time between February and April depending on the timing of the high spring tides and the river temperature, to migrate into rivers to breed. They formerly came into several of the Solway's Scottish rivers – the Cree, Fleet, Dee, Lochar, Annan, Nith and Esk – but now the River Cree is the only site, and one of only three rivers in the whole of Scotland where spawning takes place (the others are the Forth and Tay). It is thought that several factors such as pollution, blockage by dams or weirs and overfishing have all caused the decline. The fish have no longer been detected entering and spawning in the Solway's English tributaries, either – but as a friend of mine in the Environment Agency (EA) says, 'That's not to say they don't!' And small numbers of fish are still there in the Firth – picked up in seine nets during surveys by the EA, or occasionally caught on rod-and-line, or in haaf-nets, or as accidental by-catch on the shrimp boats.

A new Marine Conservation Zone around Rockcliffe Marsh and the mouth of the River Eden was designated in 2019, specifically for protecting sparling, in the hope that they might return in sufficient numbers to breed – but it's likely that the fish will need some positive help in the form of reintroductions. The Galloway Fisheries Trust's (GFT) two-year project monitors spawning and gathers data about the fish on the Cree, and has been important in finding out more about the biology of this elusive and rare fish. Apparently, one of the hints that the fish are on their way upstream is the sudden increase in the number of predators such as cormorants, goosanders and herons along the banks (it is uncanny how the birds seem to know – I have seen a heron waiting by our pond for two days before the frogs returned to spawn).

Attractive greenish-brown fish, about thirty centimetres long, they have a characteristic smell – of cucumber. But a friend who has seen the incoming shoals in the River Cree, is less certain. 'It's not exactly cucumber. But the air over the river is different, there's a definite scent of something herby, or perhaps like violets.' A miasma of exuded signals.

As the females lay their eggs amongst gravelly rills the males release clouds of sperm; the white eggs are sticky and large clumps adhere to gravel and riverine vegetation. For a few hours each night, over a period of four to five days, the fish spawn – and then they return to the sea with the river current and the ebbing tide, or die. That sparling influx into the Cree provokes strong memories and stories amongst local people, stories that have been captured on a video for the GFT.[12] A man tells of putting out the nets, and catching 'a ton. Two men were unable to pull the net into the bank.' A woman remembers that, as recently as 1980, someone would come up from the river and say, 'Right, get your buckets, children. We'll go down to catch our tea. And there was fish right across from one side of the river to the other!'

'I know when I've caught one even before I see it,' Mark Messenger, an English haaf-netter on the Firth, told me. 'It's the smell!' A haaf-netter knows immediately when he has a catch: he up-ends his net and either kills or releases the fish. The process takes two or three minutes, no hooks have been used, the released fish – sparling are protected – is unharmed and able to swim away at once.

'It's one of the last wildernesses in the country, the Solway Firth. And I think haaf-netting is probably the best excuse to go and stand out in that wilderness,' says haaf-netter Mark Graham on BBC Radio 4's *Open Country*.[13] Haaf-netting was introduced by the Vikings, presumably after they invaded the Solway shores from their base in Ireland. The haaf is large and heavy, its beam of about five metres, supposedly as long as a Viking oar, and it has three uprights attached – the long central pole acting as the handle – between which the net hangs loosely like a poke or pocket. Most of the netters make their own haafs, usually now of pine or 'greenheart' rather than pitch-pine, and they tend now to buy their nets rather than knit them. Haaf-netters may catch a range of sea fish like grey mullet, plaice and sea bass – and even catch-and-release the occasional protected spar-ling – but it is the catching of salmon and sea trout that is their *raison d'être*.

I'm fishing 'on the back of' Mark Messenger's licence from the Environmental Agency and now, kitted up in fleeces and waders, feel-ing as round and fat as the Michelin man, I follow him across the saltmarsh on the west of the viaduct embankment at Bowness and across the surprisingly firm mudflat to where oystercatchers are scur-rying and probing at the water's edge. Mark carries the haaf and we keep on walking, straight out into the water. The tide is still on the ebb, and as we wade out Mark warns me, 'You always need to keep an eye on where you are, look back at the shore', because on a big tide like this, the deep creeks in the marsh could well be hidden by the water when we head back. Testing the depth ahead of us by dipping the beam-end, Mark decides we'll fish a 'hole' where the water is deeper, and as we talk I learn new terms like 'drops', where the sand banks up and spills over the edge, and fine Nordic-sounding names like 'briest' (bank) and 'reestings' (standing waves). We stand waist-deep near the drop and the net drifts then tautens, as the current swirls around the hole. Some of the holes have names like Merry-go-round and Killing Hole: 'Think of it like a goldfish in a bowl,' Mark explains. 'The fish swim around the edges.'

Over on the Scottish side a group of haaf-netters from Annan are also out in the water. 'They're fishing a hem,' Mark says. 'If you think of two currents meeting, one flows on top of the other and then it drops down—' he shows me with his hands. 'And where it drops the water's flowing faster and harder. The fish can get trapped. But it's hard to fish.' The men clearly aren't having any luck, they are moving round the hem, and eventually wade away to look for better water. Soon there are signs that the tide is on the turn, for the net is drifting, and we can also begin to feel the 'stem' of the tide – as the flood-tide meets the outgoing water, the level and pressure increases. It's time to get out of the water, and as we wade back to the sand we see that other netters have come down to the shore and are waiting. The Scottish fishers too are back onshore. I become aware of a distant roaring sound that isn't just the wind, and see a line of black tipped with white out in the channel. The tidal bore is racing in; it isn't high but it is loud, and after it passes the sea rapidly pours into depressions that I had scarcely noticed, and creeps very fast across the mud and sand towards us. The silent rise of this brown-frothed edge is slightly menacing.

But there's a saying, 'If your net's in the water you're in with a chance.' There are now eight of us, and it is time to draw for positions in the 'boak' (or 'back', on the Annan side), the line of nets. 'We're very democratic,' Mark says, and I hear this stress on democracy several times. One of the netters is nominated 'baggie'– dating from the time when the lots were drawn from a bag – and will do the draw: he moves away and stands with his back to us. The others bring out their 'priests' or mells, the small truncheons that are used to hit a fish on the head to kill it, and push them upright into the sand, then (because the baggie would likely recognise whose stick is whose) each man chooses a different stick as his lot. The baggie comes back, and pushes over the sticks, giving each a number. Whoever is 'number 1' can pick his position in the boak. The men sling their bags over their shoulders, take up the haafs, and we wade out into the stream. Initially, Mark and I are at the landward end of the boak, for which I am glad because suddenly the current is very strong, the bore is roaring again, and everyone very rapidly up-ends their nets and hurries back to the sand. There's laughter and joking, 'Whose bright idea was that, then?' and we wait a while, watching the reestings and the rushing tide. But soon we're back in the water, where the level is rising quickly; as the water reaches his chest each netter peels off from the seaward side of the line, back towards the land. There is a flurry as someone nets a salmon; he up-ends his haaf, and after killing the fish with his felling stick, pushes it into his bag. Clumps of weed trapped in the net are shaken free in a shower of sandy fronds.

Criffel has vanished, blotted out by the rain that is sweeping towards us across the Firth. We are standing in a row facing the incoming tide, and there's not much talking, just an occasional comment or joke amongst the men. Mark Messenger and I are now at the seaward end of the row, and the fast-moving brown water is rising quickly, and is soon well above my waist. The surge of the tide sucks the sand from beneath my feet, and my face is salty and wet from rain and spray. The haaf-net streams out behind us in the current, and Mark and I each have our fingers looped in its upper mesh, waiting for the tug or tremor that says we've trapped a salmon. Mark suddenly up-ends the haaf, swinging the frame up and out of the water to make a pocket of the net – there's a small flounder, which he releases. I hadn't felt a thing, other than the tension of the tide. The water is chest-high now and the press of the water is strong. We must move. We wade slowly across the current to the landward end of the row. Mark is carrying the haaf and I wish I had the extra weight as I feel light enough to be knocked over and swept away with every step. For a while now we alternate between standing with the haaf pressed against us by the tide, then pushing across to new positions, the line re-forming constantly, rain and spray in our faces. But I'm dry inside my waders and waterproofs; I'm warm in my thermals and layers of fleeces – and it's exhilarating to be *in* the Solway, not observing from the margins. I am feeling the complexity of the flowing tide and what it bears, trying to read the waters like a haaf-netter – and a fish.

That was nearly ten years ago. At that time I also met Tom Dias, who had started haaf-netting in the 1980s; a haaf-net is often seen propped against the outer wall of his house. The Solway Band have a song about him:

> Tom goes fishing in the midday sun,
> Keeps on fishing 'til the ebb is done.
> In the reestings he goes out
> He goes fishing in the reestings while no one's about.[14]

Tom told me that they used to be able to fish for six months, from mid-February to early September, with no restriction on the hours, so everyone who wanted to net could find a time and a tide to suit. There were about 230–250 licences issued each year on the English side, he told me: 'When I first started it was normal working-class fellas, ambulance men, railwaymen, fellas who worked shifts, young lads.' Large numbers of salmon were caught; there are stories of well-stocked freezers. But because the salmon numbers were falling – everywhere – and there was a legal obligation to conserve the stocks, the Environment Agency introduced new rules, reducing the season and the hours for haaf-netting. 'Now it's too difficult for people to get the time. There're only about fifty licences taken out, we can only work three months of the year and now only half of that, the twelve-hour day. In other words, one-fifth of the men, working only a quarter of the time, just scratching a living.'

The number of salmon each licence-holder can keep was reduced to ten, and then more recently to three – and as I write it is zero. All salmon, in the estuaries and the rivers, are 'catch-and-return'. This is true for haaf-netters both sides of the Solway, including the Annan Common Good haafers. An undated hand-drawn map shown to me by Nic Coombey of the Solway Firth Partnership shows the position of twenty-nine 'fixed engines' – both poke-nets and stake-nets – on the Scottish side, but the use of fixed engines is no longer permitted in the Firth. On both sides of the Solway there is a real worry that these traditional methods of catching salmon and sea trout will soon die out – but the practitioners are not all 'old men', and Annan Common Good have recently celebrated haaf-netting in an exhibition and a very lively video.[15]

At the time I joined Mark in the water he owned The Highland Laddie pub at Glasson on the Solway (now sadly closed), and he ran a 'haaf-netting experience' business. After my own experience, I sat in the warm and welcoming bar, its walls covered with watercolours and oil paintings of haaf-netters and dogs, the shelves in the pool-room bright with books. The blackboard said *Specials: freshly-caught haafnet*

Solway salmon. Some locals were already seated around a table, a few sodden lycra-clad cyclists were piling their helmets in a corner and Mark, calm and smiling, told three dripping girls to drape their wet clothes and rucksacks in the pool-room. They padded around on the carpet in their wet socks and without hesitation ordered 'specials'. They chose well: crisp golden chips, salad and large hot chunks of wild salmon, with a taste and texture better than any I had eaten before. But there will be no more haaf-netted salmon on any menus unless the EA's next review in 2023 shows that the UK's breeding salmon population has increased.

So why are there now such tight restrictions on the catching of salmon, not just both sides of the Solway but elsewhere? Salmon numbers have plummeted nationally, for reasons that are not completely understood, so the fish must be protected, 'conserved'. One of the fishermen recorded in *Ebb and Flow*, a booklet capturing memories of Annan's salmon- and shrimp-fishers, says that in 1964 there were 30,000 salmon caught near Annan; even allowing for exaggeration, that is a shockingly high number.[16] The environment, conservation – these are highly politicised issues, with complex networks of interests. There are several players here on the Firth: the haaf-netters; the riparian rod-and-line fishermen and landowners; the Environment Agency and the Scottish Environmental Protection Agency; and the European Union, through its Habitats Directive (for example, salmon are specifically mentioned in the Special Area of Conservation that includes the River Eden). As the Solway Band sing in *The Haafnetters*:

> We've got rich men against us, we're taken to court,
> But the freedom to fish is a freedom hard bought.
> The government's throwing our freedom away,
> But we'll fight to the end for our way.[17]

Now, too, because of the climate crisis, with warming seas and rising sea levels, the future is even more uncertain. We can work out compromises between the various human interests, we can try to protect the spawning grounds and the estuaries – and we can hope it is not too late to help the sparling and the salmon. For – lest we forget – the main players, who need to be in this to win, are of course those charismatic, well-travelled animals, the returning salmon, both the adults and the younger grilse.

*

I am rather fond of Frank Buckland (1826–80), the son of natural historian and palaeontologist Dean Buckland of Westminster Abbey (see also Chapter 6); a big man, broad-shouldered and with a large nose and bushy beard, he was known to be a gentle and hard-working man, with a wide range of interests and enthusiasms. In 1867 he became an Inspector of Fisheries, at a time when, according to his brother-in-law and biographer George Bompas, 'At the river's mouth, and on the higher banks, men strove to take every fish which entered the river, with little forethought for the preservation of the fishery.' In 1860, 'the decline of the English salmon fisheries had become so notorious that a Royal Commission was appointed to enquire into its cause, and devise a remedy'.[18]

Through lectures, reports, repeated visits, research and even hatching salmon eggs in his kitchen, Frank Buckland persuaded landowners and authorities to remove weirs and watermills, build fish-ladders and otherwise remove 'obstructions formidable in number and formidable in construction' to free the passage of young salmon to the sea, and of adults to the rivers to spawn. On a visit to Maryport in 1867 he found many impediments along the River Ellen, including a weir near the river mouth. The major landowner came in for criticism: 'Now Mr Senhouse . . . has the key of the front door; the upper proprietors have the bedrooms and dressing-rooms and attics. If you do not unlock the front door, it is of no earthly use opening the other doors. The fish must be got over the weir at the entrance of the river. It must be done.'[19]

He also noted 'strange customs' in the estuary of the Eden in Cumberland, where 'the salmon are taken by heave-nets and hand-nets, like gigantic shrimp nets, the large ones 19 feet, the smaller 6 feet wide', and describes the setting–up of the draw (which he called 'mills').

Haaf-netting has a long tradition on both sides of the Solway, some of which are celebrated in song as well as stories. One hot and joyous evening at The Highland Laddie, the Solway Band came to join with musicians from Pentabus Theatre to play and sing songs about the Firth, focusing on the ballads from their collection, *The Haaf-netters*. The subjects of a couple of songs were known to some of the people there, and many stories were told, and photos taken and texted. After we left the pub we drove home through the summer evening, the late-night sky streaked with silky turquoise and lemon light; water in the saltmarsh dubs was silver, a heron gathered in its neck and rose from the muddy margin, *craak*-ing like Sinbad's rocs or Tolkien's orcs, and the dark rectangles of haaf-net frames were propped against the top tier of the marsh.

*

'I came to the Essex [Esk] which is very broad and hazardous to crosse even when the tyde is out . . . made me take a good Guide which carry'd me aboute and a crosse some part of it here and some part in another place it being deep in the channell where I did cross, which was in sight of the mouth of the river that runs in the sea . . . Thence I went into Scotland by the river Serke [Sark] which is also flowed by the sea but in summer tyme is not soe deep, but can be pass'd over tho' pretty deep but narrow.'[20]

So wrote the intrepid, but often scathing, Celia Fiennes (1662–1741), who travelled on horseback throughout England and Scotland, record-ing details of food and the countryside and its inhabitants as she went. On this particular occasion, in 1698, and finding nowhere suitable to sleep and nothing she could bear to eat ('clapt oat bread'!), she bought a salmon instead and crossed back into England again via the Sark and Esk crossings, noting, 'there I saw the common people . . . take off their shooes and holding up their cloathes wade through the rivers when the tide was out: but this is their constant way of travelling from one place to another, if any river to pass they make no use of bridges and have not many'.

Miss Fiennes and the 'common people' were using the waths, or crossing places, on the tributaries of the Solway; Edward I's army and victuallers, the cattle thieves or Border reivers, messengers and smugglers all used these routes across the tidal mouths of the rivers – with a careful eye on the tide! *Sol* – mud, in Anglo-Saxon and Norse; *wath, vath* – ford, in Anglo-Saxon and Norse: that great nineteenth-century chronicler of the Solway, George Neilson, argues convincingly in his *Annals of the Solway until AD 1307* that the original Sulwath, the muddy ford, was the route across the Esk, and Brian Blake thinks it later gave its name to the whole larger estuary. The purpose of the granite Lochmabenstane (which is not at Lochmaben), a large glacial erratic that was once part of a stone circle dating from about 3000 BC, might have been to mark the western end of the Sulwath, and was perhaps also important as a meeting place and to mark the Border; but opinions differ. When Celia Fiennes crossed the Esk the wath was near the mouth of the river and in 'sight of the sea' but since then, as Neilson showed in 1898, 200 years after Celia's journey, by comparing maps over a 300-year period, Rockcliffe Marsh had accreted and grown far out into the Firth (Chapter 4); Sulwath is now probably further to the east, somewhere near Metal Bridge, which carries the never-ending stream of cars and lorries across the Border near Gretna.

There are many interesting, often horrifying, stories in Neilson's paper about the to-ing and fro-ing of English and Scottish troops across Sulwath in the thirteenth and fourteenth centuries. In February 1216 followers of the Scottish king Alexander II, laden with spoils from pillaging Holme Cultram Abbey, were crossing the ford on the Eden when the incoming tidal bore overtook and drowned 1,900 men. And the comings and goings intensified during the War of Scottish Independence: in 1300, after 6,000 of Edward I's troops had waded across Sulwath and thence to Dumfries and onwards to lay siege to Caerlaverock Castle, the army's 'Victuals, in course of being taken with two carts and seven horses to Lochmaben, were carried off by the Scots "in the passage at Sulwath". The incident is typical of the guerrilla warfare pursued by the Scottish army.' Several years later, when he was old and ill – probably of dysentery – and threatened by the activities of (now king) Robert the Bruce, Edward and his army headed once more towards Scotland. But at Burgh-by-Sands on 7 July 1307, before he could even be carried across the Sulwath (such was his determination to confront his enemy once more), he was dead. His body was laid to rest in St Michael's Church at Burgh for several days before being carried to Walthamstow Abbey and finally buried in Westminster Abbey. Only a few years later, in 1319, Robert the Bruce's army came across the water to mount a vicious attack on Holme Cultram Abbey, even though Bruce's father was buried there.

Sulwath over the Esk; Rockcliffe Wath, Stonywath and the Peatwath near the Eden's mouth; the small Loan Wath over the Sark: routes within or between the two countries, according to where the Border lay. During the existence of the Debatable Lands between the rivers Sark and Esk – the area between Scotland and England that was continually squabbled over until it was officially divided between the two countries in 1552 – the lawless and often violent clans of the Border reivers used the waths to mount raids on their neighbours. But there are also two much bigger, riskier crossings across the Firth itself – Sandywath, from Dornock to Drumburgh, and the Bowness Wath, from Bowness to Annan Waterfoot. These crossings are not lines drawn in the sand and seabed, they are necessarily approximations, for the channels and sandbanks can change even within a day.

Back in the summer of 2015 I interrupted my bike ride around the pleasantly level and quiet lanes of the Upper Solway to have lunch at The Highland Laddie at Glasson. Mark Messenger chatted about the changes that had recently occurred in the upper reaches of the Firth. Two big rocks, the Altar Stane and the Drawing Stane, had recently reappeared at low tide off Annan Waterfoot after years of being hidden beneath the sand. The Altar Stane is another of those glacial erratic boulders, like those on the Allonby shore, that has been given a name, for it marks the southern boundary of the Royal Burgh of Annan and is also a boundary for fishing-rights. Mark said that it was now possible to cross the Firth just west of the remains of the Bowness–Annan viaduct: 'You can wade across, there's that much sand. I've never seen it like that in eighteen years.' Who could resist such an opportunity?

However, when I met him a couple of days later he told me that the situation west of the viaduct embankment had already changed: a deep channel had opened on the far side and we would no longer be able to cross there but would have to try elsewhere.

We drove down the coast road in Mark's van and at Bowness we pulled in behind cars already parked on the roadside. Several men were wandering around, chatting and pulling on waders and yellow water-proofs. Haaf-nets were propped against the bank. Mark nodded towards a man who was taking waders out of his car. 'He's the one who went for a swim last week,' he grinned. The fisherman (we'll call him Jim) got knocked off his feet and was swept a short way downstream into a 'hole'. 'I couldn't get out,' Jim told me. 'I was trapped in it. Then the tide turned and washed me out. I was only in for fifteen minutes but it was cold.' Nevertheless, here was Jim, back again, kitting up, pulling the strap of his bag over his head, readying himself for another few hours' fishing in the Firth.

Downstream at Silloth, the tide would be on the turn and would soon be flooding back up the Firth but here, just upstream of Bowness-on-Solway, the tide was still on the ebb. There is a long period of slack water between the ebb and flow; as Mark said, 'We don't get twenty-four-hour tides here.' Sandbanks and mudflats stretched out from both the English and Scottish shores and in places sandy shoals were begin-ning to show beneath the surface of the water. It looked as though it should be an easy crossing, but to reach the water we would have to cross slippery mud and stones, and then a long stretch of sand. 'It's hard to believe it's changed so much in just a few days,' Mark said. 'This was all rocks last week, like that over there', pointing to an emerging scaur. Now, mud, grass and a tangle of river-borne twigs and leaves had been pushed right up to the bank along the margin of the road. He pointed out curving lines of débris on the shore, which showed how the current of the flood-tide came close to the Bowness side as it swept round the corner. Over to the north-east the River Esk swirled round the far point, but instead of immediately joining the flow of the Eden it had, in the past few days, carved itself a new channel close to the Scottish shore. An aerial view from a gyroplane or a satellite view on Google maps shows the clearly braided channels of the Eden and the Esk; it seems counter-intuitive that rivers would meander widely across the bed of an estuary, especially when two large outflows are competing for the space, but the Esk seems always to have been particularly adventurous. A comparison over time of the Esk's channels, drawn from maps from the mid-1800s to the present, points up the quite extraordinary changes in the Esk's path: in 1857 it pushed south and down along the English shore but four years later was tucked up around Redkirk Point and Scotland; in 1946, it flowed south to the north side of Rockcliffe Marsh then made a slight northwards loop towards the centre of the Firth, and by 1963 it was almost back in its more northerly 1861 channel, where it seems to have approximately stayed.[21]

Mark had also been chatting to Billy – another haaf-netter who was getting kitted up – about the state of the channels and the shoals. Billy had been out fishing this stretch of the Firth the previous day, and he pointed out a white house we should aim for on the Scottish side, and where standing waves indicated a shoal. 'Why d'you want to do this?' He was laughing at me as I was flapping the over-large top of the waders Mark had lent me. 'Something you've always wanted to do, is it?' I agreed but admitted I was slightly nervous, now that I'd seen how far we had to walk and wade (in both directions – there would be no waiting taxi on the Scottish side to bring us home via Gretna!). 'You'll be okay. It'll be easier coming back. And Mark'll look after you, won't you, Mark?' I knew he would. He constantly scans the surface of the water, and explains what he's looking for and what it means. If you want to cross the Solway, on one of the ancient 'waths' or anywhere else, you couldn't do better than ask a haaf-netter. Mark's own haaf-net was on the marsh the other side of Bowness, so he borrowed one that was propped against the bank. He said we'd 'have a look and see what we can find' – but he also told me later that carrying the net was useful in case 'anything goes wrong', because the frame can be used as an anchor or support.

The weather was good for a crossing: very little wind, the air warm despite the grey sky, the waves small, so we waded out into the shallow water. Soon it was knee-deep, and then above the knee. Walking wasn't difficult. But then we reached the current, where the Eden – busily carrying water that had recently fallen on the fells – was making its presence felt. Walking against the current became slower and harder. Whenever I stopped, the ground beneath my feet was sucked away. Occasionally, a sudden stronger swirl caused me to lurch. I began to feel hot and sweaty from the effort, and anxious about having to make the return journey, but I was determined not to fail, to keep going and not show how weak my legs suddenly felt. Mark waded slowly, keeping an eye on me while pretending not to do so, occasionally asking if I was all right, suggesting I could hold the bar of the net if I needed. He also stopped frequently to look at the water. 'If there're salmon, they'll be coming down like bullets,' he said. 'Look out for flashes of white.' And, 'We're heading for where the water's smoother, over there. Another fifty yards and we'll be over the worst bit.' And so we were, eventually, though there was still a fair way to wade through the shallower, calmer water, towards a sandbank splattered white by resting gulls and oystercatchers, and decorated with the small pink shells of tellins. Along the coast in front of us were rectangular box-like buildings, their sickly green paint failing to blend with the surrounding trees and thickets. These were the munitions stores of Central Ammunition Depot (CAD) Eastriggs, the remaining section of CAD Longtown.

Eastriggs was part of HM Factory, Gretna, the UK's largest factory for the production of cordite, built in 1915 and covering an area from Longtown in England to Dornock near Annan. Ten thousand navvies were employed to rapidly construct the four separate production sites and the two townships to house the workers, so that cordite could quickly be produced to overcome the shortage of shells experienced by the British soldiers fighting in France. (As with the construction of the Solway Junction Railway, there is no mention of a navvies' camp.) The story of cordite production is told in an exhibition in the fine new building of the Devil's Porridge Museum at Eastriggs – the name comes from a comment by Sir Arthur Conan Doyle who, as War Correspondent, visited the factory in 1916 and watched the women kneading together the mixture of gun cotton and nitroglycerine 'into a sort of devil's porridge'. Nearly 12,000 women worked at Gretna, making this very volatile paste and then turning it into the cordite propellant that was later packed into shells. As photographs and accounts in the museum show, the chemical fumes often turned their skin yellow, so that they were known as the 'canary girls'; they lived in hostels that were run by matrons; their overalls had no metal buttons in case a spark led to an explosion – but apparently there were compensations, for they were well paid, and 'gained a sense of freedom and independence for the first time'. At Gretna and Eastriggs there were cinemas for silent movies, dance-halls, reading rooms and tennis courts where the women – for female workers outnumbered males by two to one – could spend their money and enjoy themselves, yellow complexions and all. By 1918 the factory was producing 800 tons of cordite a week, which was taken by rail to Dover and shipped to Normandy – where it was used to kill tens of thousands of German soldiers.

Mark and I ambled about on the sand and in the shallows, looking at the view, admiring Bowness as seen from the Scottish shore, watching the other haaf-netters downstream on both the English and Scottish sides. Mark had a Scottish licence as well as an English one, but we had set foot on the Duke of Buccleuch's territory, where haaf-netting wasn't permitted. Our crossing had taken just over twenty minutes and now, having examined Scotland and rested, it was time to head back to England. In the forty minutes or so since we walked onto the sand at Bowness, the water level had dropped even more and new sand was exposed. We were also walking with, rather than against, the rivers' flow so the going was easier. I wondered if we were now wading through fresh water rather than brackish but it was so turbid and brown with silt that I didn't feel inclined to taste it. Ahead of us were reestings – standing waves, occasionally, chaotically, tipped with white – an indication that the water ran fast and shallow over shoals. It made interesting wading, to feel the sudden change in speed and force, to stand and watch how the patterns and heights of the waves changed subtly, to wonder how a streamlined fish like a salmonid or sparling would use the tide. Rays and flounders would respond to quite different signals, as they rippled low over the sand and mud like stealth-bombers, looking for prey. Now would be a risky time for mudshrimps and snails to be exposed on the mud's surface, but the water was so silt-laden that even with a glass-bottomed instrument like a bathyscope it would be impossible to spy on their activities as the sea returned. The flood-tide was starting to reach Bowness, the haaf-netters would be feeling the stem, that change in pressure when the opposing waters meet, and so we needed to return to the English shore. There, the ground underfoot and the air around us were disappointingly predictable. I had experienced a little of what the drovers, soldiers and raiders – and surely also ordinary people just needing to cross to the other side – must have experienced, but I had good waterproof clothing, I had not been driving a herd of cattle, nor had I been weighed down with armour or heavy equipment for going to war. It was good to be dry and warm, and going to the pub for lunch.

Mark had pointed out the lump of the re-emerged Altar Stane further to the west, too far for us to reach on that tide. Every July, Annan's Riding of the Marches celebrates the creation of the Royal Burgh by riding horses along the town's boundaries and, if the tides are right and the stone is visible, riding out to the Altar Stane too. According to haaf-netters who have visited the Stane, it is inscribed with initials and various names, some of them scratched into the rock, others painted; it would take some effort to trudge across the mud with a paint-pot and brush.

What has not re-emerged from the bed of the Firth are the bells from St Michael's Church in Bowness. The story is that they were stolen in 1626 during a raid by the Scots but were dropped overboard, whether accidentally or on purpose isn't clear – the tales vary – in what is now known as Bell Dub or Bell Pool. The Bowness people took their revenge by crossing the Firth and stealing bells from Dornock and nearby Middlebie. The stolen bells are still there, resting on the floor inside St Michael's – knee-high, dull grey and heavy. It became a tradition that every new minister at Dornock should write to Bowness asking for the bells' return, but he has always been ignored or refused. There are still re-enactments of the raid and in the summer of 2019 the plan was for an Annan-built skiff to be rowed across to Bowness for this very purpose, but the big tides and rough weather caused the event to be cancelled. With all the activity at the waths, it would not be surprising if a plethora of dropped and discarded objects emerged, but the changing profiles of the channels and mudflats, and no doubt a reluctance to investigate in such dangerous territory, have meant that little has been found.

This had been a very different experience from when I went haaf-netting: then I had had to wade out into the flood-tide, standing with the net as the water rose to chest-level. I'd learnt about briests and reestings, hems and holes, seen and heard the bore, watched as the water crept so unnervingly quietly into the channels on the mudflats. That had been about the power and the vagaries of behaviour of the sea. But today had been about the power of the rivers that modify the inner Firth, on their own and during their interactions with the sea. I'd learnt that every time you go out into the Firth you not only have to see but also to understand the subtle clues that tell you where it is safe to walk or wait, where shoals have formed or holes appeared and, unlike the creatures that live within the water, you have to do so from above the plane of their existence.

*

Nellie and Pintle, High Netherma and Maston, Metalstones, Archie and Popple scaurs: 'The names go back a terrible long time,' Ronnie

Porter said. The names of the stones are part of the oral tradition of the Allonby shore, and neither Ronnie nor his wife knew how they should be written. There's been a long tradition here of catching fish and shellfish, with fixed nets, fixed and baited lines, fixed lobster pots, or with handheld nets or rods or rakes. The boulders and scaurs – reminders of vanished glaciers – are markers for the fishing grounds, and Ronnie is one of the very few people remaining who know their names. Still spry in his early eighties, and a kind and gentle man, he walks on the shore most days and only gave up shrimping with his heavy 'shoo-net' a couple of years ago. 'But it wouldn't be any good now anyway. I used to go down below Popple, but the sand has gone, it's all gravel. You can't use a net there. Though it could all change in the next tide, any road.' For the shore profile is constantly changing; the spring tides and the storms shift the sand, revealing or hiding the scaurs; sometimes the big boulders are exposed to their bases, at other times only their tops may show. Below the village it's obvious, as Ronnie says, that 'the sand has gone off the shore, there are lots of extra scaurs. The gravel bank is moving north-west, and there's a hollow behind it.' A few years back he found a row of newly exposed red clay fishing weights lying in the gravel.

I first met him about five years ago, when the low-tide conditions were perfect for our proposed evening walk, the water calm and long soft shadows accentuating the rocks and rippled sand. I'd already noticed that the rocks seemed to shape-shift and change in size depending on my vantage point: looking from the dunes or the road, I'd thought I understood their topography, yet from a hollow on the shore, the scaurs merged confusingly, only to reappear as individuals when I walked out onto the domes of sand that formed stationary waves along the shore. But Ronnie pointed out the obvious markers such as the huge square Maston, down below the end of the Edderstone Road and inshore from the scaur called Archie; and Hanging Stone at the edge of Dubmill Scaur. It was around Maston, he said, that they would put out their long-lines for cod, baited with shrimps. The lines, with hooks about six inches apart, were tied between metal stakes, and codling – 'loads and loads' when Ronnie was young – were caught, then retrieved when the tide went out. His grandfather would attach prices, and Ronnie and his friends would set off on their bikes to sell them, cycling round the local farms and villages with the strings of fish dangling from their handlebars. In early spring, he said, they would put out long-lines for skate, baited with chunks of herring. The metal stakes are still there, at Maston and elsewhere on the shore, sometimes hidden, sometimes revealed by the shifting sand. I stood still and looked back towards the village, towards Crosscanonby and the mile-fortlet on its little hill, and I repeated the names Ronnie had told me, trying to fix them in my mind.

Anchor Scaur.

Popple – below the village, and which is increasingly being exposed.

The two boulders, High and Low Netherma (because they're visible at high and low water).

Then the scaur called Metalstones

'What's next?' I asked Ronnie. 'Is that the one you call Archie?'

'No, that's Matta.' It's slightly higher than Metalstones; the reefs forms ridges.

And then we crossed onto Archie, the biggest scaur, and there were two large boulders, Nellie and Pintle. At least, Ronnie thought they were Nellie and Pintle, but then as he looked north towards Dubmill, he was not certain. 'I'm not sure any more,' he said. 'Maybe that's Nellie up there.' We worked our way through the 'coral', as the honey-comb-worm reef is called locally, and reached the far edge of Archie, next to a broad swathe of firm and rippled sand. Stretching way out into the Firth were two long lines of rock, Hill ('it's difficult to get to, and you don't have much time there') and Far Hill. But we could cross the sand to The Squash at the edge of Dubmill Scaur, and on the way I patted the great Hanging Stone, which I myself so often used as a marker for my guided 'Dubmill shore-walk'. This time, Hanging Stone was standing proud and fully exposed, its top scarfed in green weed; sometimes, when the sand banks up around it, it is diminished and I am disorientated. There were many shallow pools in The Squash, in which brown fronds of oarweed were growing, smooth *Laminaria digitata* and the crinkle-edged *L. saccharina*. Ronnie told me this would be a good place for prawns: 'They hide under the weed. I'll have to come back here!' He told me how on a sunny morning the prawns would be in amongst the kelp and under rocks. 'I'd have to howk them out with the click into the net.' At night they glowed in the light of a torch. Later he showed me the hand-net, much smaller than that used for shrimping, and the click, a stick with a metal hook.

We stood quietly for a while to enjoy the peace and soft light of the evening, listening to a curlew and oystercatchers; watching a heron wade into a pool up to its chest then pose, motionless, neck curved. Then, from Dubmill, Ronnie pointed out The Shotts, Number 1 and Number 2; and Crinla Lake and the Coving Stone, with Whitestone Gate (marked by a whitish boulder); and Point of Tail scaur, which goes out into the Firth from Crinla. 'Some people also call The Shotts 'The Roads'. They're like straight roads going out. The old tale is that they were done by the Romans. The coral was cleared, the stones cleared. They'd throw the big stones out to the sides.' I wondered if the reason had been to make safe passage for boats to come up the shore, but Ronnie thought the 'roads' were for catching fish. 'You'd put baited lines across – the cod would swim up and get caught.'

I asked him too about the rows and triangle of stones a little further north on the Mawbray Banks shore, but he didn't know what they were for. There are lines, sometimes tens of metres long, straight, right-angles and triangles: boulders large and small, some at least a metre across and high, in places encrusted with barnacles and mussel spat, elsewhere scoured clean. The ground inside the triangle and the right-angle's elbow is sandy, and pooled with water. A year later, at low tide, I flew above the shore in a gyroplane and marvelled at the straightness, as though the marks had been ruled with a crumbling stick of charcoal. On the ground the jumbled pebbles of the rocky scaurs make no sense, but from above it is as though they too have been smeared – by the broad brush of the sea – into stippled bands at an angle incident to the shore: the boulder lines, superimposed, now make sense as barriers, aligned to trap fish in the ebbing tide. How old are the traps, if that is what they are, and who struggled to move those boulders and ensured the lines were straight? In the Beauly Firth are very similar structures, known to be fish traps – in Scotland they are called yairs.[22] But here they are part of the oral history of Solway fishing that has been forgotten.

It was amongst the *Laminaria* fronds in a pool amongst the *Sabellaria* reefs that Ronnie had said he would catch prawns. Now, seeking stories of shrimps and herring, I visit Ronnie and his wife once more. Apparently, the best places to catch shrimps had been near Netherma and High Netherma. 'You catch shrimps on the low tide, sometimes three-quarter hour before that and a little bit after. Sometimes when it's half-sand [i.e. half-tide] you can get them too. So we'd keep an eye out for Netherma and once it was showing, we'd be away down with the shrimp nets. We'd start there and work over East Sand towards Dubmill Scaur.' Other useful places were on the scaurs of Metal Stones, Matta and Archie, 'but they're no good now cos lots of the sand has been washed off and it's all cobbles right now'.

He takes me out into the back courtyard, the former yard of the herring shed, and shows me the shrimping net. I'm astonished by its size; the handle is about eight feet long, and the radius of the loose pocket of semi-circular net is about five feet. The leading edge is a straight board, to which the handle is bolted by a broad iron strap, to give it weight. 'The old chaps, their nets were seven foot wide. Mine was six foot six but I had to put a new sandboard on about twenty years ago, it's a bit of hard wood, a bit of mahogany, and it was only six feet.' Ronnie laughs. 'It's about all I can manage these days.' He explains that the shrimps come onto the shore with the incoming tide and then bury themselves in the sand. 'But where there are little pools, they'll be there all the time. You're pushing the net along, water's sometimes up to here [he points to his ankle], sometimes deeper [knee]. I've caught big plaice, a full-size turbot, even, right on the edge of the sea.' At Allonby, shrimpers push the net in front of them: 'It's called a shoo-net.' At Maryport, though, they pull the net behind them: 'They have to pull it onto the sand to sort and empty, it's all mixed up with gravel.' He smiles. 'We've been doing it much longer at Allonby, about four hundred years, like!'

He shows me how to push the net – hands holding the end of the pole, and pressed against the belly. I try it, and decide I would probably have given up within fifteen minutes! Ronnie would have a pannier strapped on his back – he prefers a plastic fishing box with holes to drain the water. 'You've got the net out front, and when you want to empty it you turn the pannier round, throw as much rubbish out as you can – there'll be crabs and weed, small plaice – and tip the rest into the pannier. I can sort one lot in the box as I push . . . They're the same as what they catch at Silloth [on the boats]. They taste better than what the Silloth shrimps do, it's what they feed on.' These brown shrimps, *Crangon crangon*, are six to seven centimetres long, and unlike the mudshrimp *Corophium* are Decapod Crustacea (see Chapter 1). The Allonby shore is sand and gravel, so the brown shrimps would not be preying on mudshrimps here; out on the banks in the channel they might be more of a threat. But the brown shrimps, unlike the tiny mudshrimps, are themselves tasty prey for humans. What sort of numbers of shrimps did Ronnie catch – hundreds?

'Oh no. *Thousands*! I'd put them in the colanders, and I'd sit at night in there – with me little dram – watch the telly and shell them. We [he nods at his wife] like them the way they are, but my mother used to pot them.'

At Silloth I walk into the shed that says 'Ray's Shrimps', and ring the bell on the tiny counter. Joseph Ray comes out from the preparation room at the back, and I buy a pot of Traditional (Spiced) Potted Shrimps, and we chat about the business. It is June, and so far there are few shrimps about, and the trips out in the Rays' boat, *Jolanda*, have not been productive. *Jolanda*, like the two other shrimp boats, is tied up in the outer harbour at the dock – the tide is out and the boats are sitting in shallow murky water that is fringed by mud. Feral pigeons are cooing on ledges by the inner dock's gates, and three crows waddle and squabble over the torn remains of a fish.

Danny Baxter's shrimp boat, *New Venture*, is a blue catamaran with an open cabin and yellow-painted superstructure and, unlike *Jolanda*, whose nets are shot from both sides, her single beam trawl is shot from the stern. Danny is a round-faced and smiling man, frank with his stories ('Fishermen are full of stories. Though we twine more than farmers – everything's got to be just right!') and we laugh a lot during our conversation. We lean on the rail by the port office, looking down at his boat as the sky darkens and thunder mutters ever closer; it is July and has been a month of weather extremes. He tells me about the boat: 'We've had her about twelve years. She was built in Wales, and we bought her for cockling [in the Solway] originally – then the fishery stopped so we switched to shrimps.' The trawl net is rolled up against the beam at the stern, and there's a system of chains and pulleys from the rigs above. I have seen the boat trawling slowly just off Silloth, its net just visible at the surface, and the stern trailed by a jostling crowd of gulls. Danny explains the beam is seven metres long and the triangular net is held against the bottom by a semi-circle of 8-inch rubber bobbins. 'You're trawling at about three knots, it's a good speed for shrimps, but any faster and you can't stop quick enough if the net gets caught – it's a twenty-five-tonne boat so takes a bit of stopping!' Mesh size is also very important, and I get more details: there's a 80-millimetre net bag in the front to stop the bigger fish getting into the main 28-millimetre net, and a 'veil' separates the smaller fish from the shrimp, so that the fish can exit out of a hole in the bottom. I ask about by-catch – and he mentions dogfish, plaice and skate. And the occasional sparling! 'My brother will say, "There's a cucumber fish in there" – the smell is really strong, you know right away.'

The black clouds are roiling above us and raindrops start spattering in the dust; suddenly the rain becomes torrential, and we dash for our vehicles and drive to Danny's house for more chat and big mugs of tea. The season for catching shrimps is usually from April to November, Danny tells me, but they try to keep going as long as they can. Last winter's catch was poor, but 'three years before, there was more in the winter than the summer. It's very variable. We've only got going in the last fortnight, I hope we have a decent second half.' He and his brother can be out for seven to eight hours, sometimes setting off at two in the morning, depending on the tides. Danny likes being on the Firth at night, likes seeing the dawn. Today they caught about fifty kilos, but in the summer they would normally be hoping for eighty to a hundred kilos in one day. But 'If we catch only twenty kilos, that keeps the lasses who's peeling going – and it gets us out of the house!'

One time, a couple of years ago, they caught more than 500 kilos. Half a tonne! Danny warned me about fishermen's tales, but I try to imagine how many animals that would have included, what percentage of the local shrimp population. A passage about a fleet of Japanese shrimpers in John Steinbeck's *Log from the Sea of Cortez* comes to mind: '. . . soon the cable drums began to turn, bringing in the heavy purse-dredge. The big scraper closed like a sack as it came up, and finally it deposited many tons of animals on the deck – tons of shrimps, but also tons of fish from many varieties'.[23] This was in 1940, when Steinbeck and Dr Ed Ricketts – the inspiration for 'Doc' in Steinbeck's novel *Cannery Row* – were on an expedition to collect marine organisms.

Those ships 'literally scraped the bottom clean', but Danny says that because nothing is fixed and growing in the sandbanks where he fishes, there is little for the trawl to damage. I'm inclined to believe this, because the topography of the sandbanks changes constantly; it would be hard for algae and animals to maintain a foothold. The shrimp boats don't put down their nets in the channels because the rocks, 'the brick' as he calls it, will tear the nets.

The shrimp are killed immediately on board in boiling water. I had seen the large cylindrical boiler on *New Venture*'s deck. 'We use sea-water for boiling and for washing – it's pumped in from underneath the boat, and we heat it with diesel. We can't carry fresh water, we'd need to use nearly 100 gallons water at a time!' The shrimps curl up when they're killed and this makes peeling easier back at the shop. If the shrimps aren't for selling peeled or potted, the Baxters freeze them in five-kilo bags, and drive a vanload down to King's Lynn for processing. I'm puzzled why the potted shrimp I buy are so small. Danny explained that the shrimps come in all sizes, but people prefer the smaller ones for potting – because then there are more shrimps to the pot!

I've been trying to understand whether these brown shrimps would prey on mudshrimps, as stated in the MARLin notes. Danny is insistent that the boats don't trawl over the mudflats because there are no brown shrimps there, so there wouldn't seem to be any overlap: *Crangon* and *Corophium* would rarely meet. And the variability in *Crangon* distribution and availability is a puzzle. 'The navigator's accurate to give or take five yards,' Danny says. 'You can set it so you go back to the same place. One day and there'll be nothing there, then next day you'll be saying "where did they all come from?" Or they'll be there at night and not in the day, or there in the day and not in the night.' Years ago, my marine biologist friend Dr Jane Lancaster spent some time trying to work out the biology and movements of the Solway's brown shrimps, but she concluded that she couldn't draw any conclusions, there are so many variables. However, I have a postscript: a few months after I met Danny, and in mid-September, when the weather had suddenly switched from a week of dark, gloomy days to bright, warm sunshine, I wandered out onto the mudflats near the viaduct embankment. The tide was so low that the distant water was invisible from my perspective on the flat shore, and I felt I could have walked dry-shod across to Scotland. There were *Corophium* burrows; mudsnails, *Hydrobia*, their movements barely perceptible, were squiggling their slow traces over the surface and on the pebbles; lugworms had ejected coils of muddy sand; and amongst the disorganised patterns of sand ripples were small pools of trapped water where little rafts of quartz grains slowly rotated in the wind. Small things skittered in those pools, moving so fast as to be almost subliminal, and then settled in the mud with a flurry. I wished I had a magnifying glass, but eventually the skitterers resolved into nearly transparent, tiny, shrimps – *Crangon*, brown shrimp youngsters, there amongst the mudshrimps.

Jane Lancaster found it was not even possible to state whether fishing for brown shrimp is sustainable. It is certainly unpredictable, and one of the Silloth shrimping families has already given up. Just across the Firth in Annan only one small boat remains, with a small trawl net ('More of a hobby, really,' someone told me), but at the end of the nineteenth century there were about fifty shrimp trawlers working out of Annan. Danny Baxter said his grandmother used to tell the story that the family had been shrimpers living around Morecambe Bay but on one trip into the Solway there was a storm that drove them up the Firth and they ended up taking shelter in Annan – which is where they stayed 'until they were driven out for being naughty, like,' Danny said, so they came across to live in Silloth instead. That Annan shrimpers originated from Morecambe Bay is part of the local history: in *Ebb and Flow*, the late John Willacy tells of his father's boat blown 'away up here, up the Solway Firth to where they did not know. Most of them found their way into Annan harbour [and] they thought it was such a good place for the fishing they decided to stop and live here . . . That's why you'll find there are Woodhouses, Woodmans, Baxters, Willacys etcetera.'[24]

The shrimp boats had a single mast and were half-decked, a design that originated at Arnside in Morecambe Bay; a photo shows a small fleet of the boats at Annan Waterfoot, their red jibs and gaff-rigged mainsails bellied in the wind. Other accounts in *Ebb and Flow* tell of the women shelling, potting or packing the shrimps, and buckets of shucked shells being carried down to the river 'in zinc baths' and dumped for the ducks and eels to eat.

Today it is hard to imagine the former busyness of Annan's quays and estuary, with its shipbuilding industry and its flotilla of shrimp trawlers, herring boats and the whammel boats used for catching salmon with drift nets. The tide is out and the narrow channel of sticky mud is bordered on one side by the tall red sandstone quay, and on the other by a high embankment from which grasses and sedges are encroaching. A decaying wooden hull is sinking into the mud, propped by an abandoned trawler that is moored against a warehouse, and a few mallards are paddling in the shallow water, one cackling hysterically like a wicked witch. A little further downstream, where there is more water, a scallop dredger is tied up against a wharf for repairs. Here on the quay, though, there are illustrated information panels telling of the harbour's history, and of the Annan Harbour Action Group's determination to partially restore the area. It is not a sad place: there is such a strong sense of how the people who lived here really knew how to work with and make a living on the Solway.

Danny offered to take me out in the *New Venture* when the weather improved – but it turned out there could be some practical difficulties: a six- to eight-hour trip, the diuretic sound of the water against the hull, no bucket, not even a curtain to hide behind . . . As I was leaving his house, I happened to mention the honeycomb-worm reefs. There is another species, *Sabellaria spinosa,* the Ross worm, that lives permanently submerged offshore; the worms are bigger because they are always underwater and so can filter-feed all the time. 'Sometimes we might get a handful in the net,' Danny said. 'We call them nar!' What sort of a word is that – nah, nar, naa? We both started laughing, and even now I can't help grinning as I say it aloud.

*

'What's the point of them?' the shore-walker had asked, about the polychaete worms.

'Amateurs are not fond of worms; nor until they have seen *Serpula, Sabella* and *Terebella* expanding and waving their beautiful tentacles in the water can they understand why we have taken so much trouble to secure them. And yet, apart from their beauty, the worms deserve our study. Their structure is full of interest.'[25]

So wrote George Henry Lewes after his rambles on the shore at Ilfracombe with George Eliot.

Are worms – and sparling, salmon and shrimps – merely 'fodder', part of the human food chain? What did I feel when I held the haaf-netted salmon? Wonder, exhilaration – but now, great sadness. Yet I still eat shrimps . . . and also marvel at the ways humans have devised means of catching marine life, with their nets, great and small. We have always treated the seas and estuaries – particularly the estuaries, for we watch the sea replenishing them every day – as a source of food, but there is now a changing mind-set from 'grab all you can, there'll be plenty more', to the realisation that 'plenty' is becoming a lost word.

9

Changing times

Moricambe, or 'Hudson', Bay is the resting place of Lockheed Hudson bomber AM 771, which crashed into the Solway Firth in 1942 (see Introduction); large numbers of these unwieldy aircraft ended up in the Firth during World War II, hence Moricambe's alternative name, and the story of these disasters has been unpicked at the Solway Aviation Museum. When I visited the museum at Carlisle Airport to find out more, a storm was approaching – the River Eden would subsequently burst its banks and flood Carlisle – the wind was scooping sheets of water from the empty runways, and the museum's aircraft had been covered with tarpaulins and fastened down with guy ropes. Leaning against the wall of the museum was a huge propellor, black and solid, its twisted metal blades pitted by marine organisms – it came from a Lancaster bomber which crashed into the sea near Millom in 1943 and which had been fished out of the Solway by a trawler off Whitehaven, fifty-two years later: a striking example of the effect of longshore drift. Newby Tate, who had offered to show me around, had to cling onto the door to stop the wind wrenching it off its hinges, but although the museum was chilly and closed for the winter, he and his wife Margaret gave me a warm welcome and a cup of tea. Newby, a kind, soft-spoken man with a twinkle in his eyes and a ruddy face, was at that time director of the Aviation Museum and Margaret was the treasurer and secretary.

In the centre of a room lined with posters and memorabilia was a glass display case, its shelves crammed with detailed and accurately hand-painted models of aircraft, each accompanied by a small numbered card. We flipped through the pages of the loose-leaf catalogue that listed aircraft from France, Germany, America, Russia and the UK, and finally found 'Lockheed Hudson' listed as no. 86. Card number 86 took some spotting amongst the wings and wheels, but we eventually saw it in the centre of the middle shelf, marking a large aeroplane with a clear dome – the gun turret – that was fitted dorsally towards the rear. The plane was a heavy-looking beast with twin propellors and two vertical fins on its tail. Newby said that when he was in the air cadets they were taken to an indoor mock-up of the Hudson's turret to try it out. 'They had silhouettes of enemy aircraft projected on the walls, so it was like being in a big amphitheatre. The gunners would have to sit in it and practise,' he told me, but this was difficult because the turret also had to be swivelled by the gunner while he trained the gun on its target. The Lockheed Hudson was 'chiefly associated with RAF Silloth', according to the notes on the wall. How many had crashed into the Solway? Anecdotal evidence has escalated the number to 'dozens', even, in one estimate, eighty-seven, and the real evidence still makes unhappy reading. The list of 'Serious crashes of Hudson Aircraft from Silloth 1940–42' totals sixty-four. Some of these 'stalled, hit trees', 'hit hill', 'overshot on landing, hit truck', 'dived into ground', 'hit mountains', but seventeen of them 'ditched', 'crashed in sea', 'stalled and spun into sea' or 'dived into sea' in the Upper Solway. Crashes into the sea, and on the ground where the aeroplane inevitably turned into a fireball, led to horrendous loss of life, and the bodies of the crewmen were rarely recovered. The Hudson was apparently a difficult aeroplane for an inexperienced crew to fly. The technicalities of its peculiarities are explained in a booklet, *RAF Silloth*, produced by members of Silloth's University of the Third Age (U3A), and include a 'slight swing to port when the engines were opened up for take-off'; the 'flaps were unusually powerful'; 'stalling had some interesting effects too'.[1] Additionally, those wartime crews certainly lacked experience, being young, often hurriedly selected, and often with minimal training on single-engined light aircraft.

Silloth airfield was an Operational Training Unit, No. 1 OTU, for the RAF, and airmen were sent to the base from many countries for training in navigation, bombing and night-flying. Franklin Zurbrigg, a Canadian in his mid-twenties, enlisted in the Royal Canadian Air Force in 1941 and less than a year later was sent to Britain, sailing in a convoy from Halifax to the Firth of Clyde. He was then sent to a temporary 'reception centre' in Bournemouth on the English south coast, and from there back north again to St Anne's-by-the Sea near Blackpool, where he received various types of further training. He finally arrived at Silloth on 23 November 1942. From the time he left home he recorded observations about his surroundings and his new job in his diary. He wrote about the colours and scenery of the countryside over which he flew, first in a twin-engined trainer and later in Lockheed Hudsons: '. . . he marveled at the "russet brown" and hilly countryside that spread out below him. He was also struck by how "long stone walls" divided the farms "into the craziest and quaintest shaped fields you could imagine",' and thought that the Lake District was a place he might like to come back to visit when the war had ended.[2] In early January 1943 he noted that he'd been out on a 'Splash exercise using side gun and turret', which probably involved firing on a target out in the sea. But on 13 January, the Hudson which he and three others were crewing stalled, crashed and burst into flames while taking off on a local night flight. All four crew members suffered multiple injuries and died before they could even be taken to the hospital. At the Causewayside cemetery near Silloth there are rows of simple white stone crosses, engraved with the names and nationalities of airmen from Australia, New Zealand, Poland and Czechoslovakia who had come to the Solway airfields for training in piloting and navigating aeroplanes, firing machine guns and dropping bombs; young men who had left their families to come to this far corner of a foreign country – and most of whom were killed in accidents, not war. Every year for Armistice Day, 11 November, red poppies are laid on the Causewayside graves. Franklin Zurbrigg lies next to another Canadian, his Warrant Officer on that final flight, F. Belanger.

Dr C.M. Johnson met with the families of these Canadian airmen to compile the McMaster Honour Roll Project, the stories of Franklin and the thirty-three other alumni of McMaster University who died in World War II: the details of the young men's early lives are especially moving. Johnson writes that he was 'presented with the proverbial cigar box of old wartime letters, written by an air force navigator. In another, a tenderly cared for binder was produced, containing descriptive, humorous, and astute letters from a young army officer serving in Italy. On other occasions I was loaned whole family archives or led to dining room or kitchen tables laden with such wartime memorabilia as diaries, postcards, photographs, flying logs, training notes, and newspaper scrapbooks.'[3]

Here on the Solway too, from the wartime airfields on both sides of the Firth from Dumfries round to Silloth, local history groups, oral history projects, videos and collections of old photographs have recorded the stories of pilots, mechanics, 'tea-ladies', airfield staff, the Royal Observer Corps and the Fleet Air Arm, to ensure they remain in the collective memory. A Solway Military Trail has also recently been developed.[4] There is great interest in projects that store and curate a community's oral and written accounts of former times, and funding can be found from sources such as the Heritage Lottery Fund. So there are exhibitions and collections about local shipbuilders and influencers at Maryport Settlement; videos and interviews with the haaf-netters of Annan Common Good; booklets and events about the former harbours and shipping at Annan; the *Remembering the Solway* oral history projects in the Bowness area; the websites and books with memories of growing up in these coastal areas – farming, fishing, quarrying; the collected memories of the *Plain People* of the Solway Plain; the Solway Coastwise project to gather the meanings of the names of rocks and caves along the Dumfries & Galloway coast – and many more.

As discovered by Kathleen Jamie[5] when she visited the archaeological excavations in the melting permafrost of Alaska, and by David Gange[6] when he visited the Comunn Eachdraidh Nis, the centre for the Ness local history society on the Isle of Lewis, there is an anxiety that the history, the meaning, even the language, of a community will be lost, and that the reasons for why things are what they are will no longer be there to inform the future. On both sides of the Solway, there are very many people who want to remember and talk about their ways of working and the places they know – with reference to ports or quarries or fishing or steel making – and so much of what they speak about relates to the past: to childhood, or to a previous generation, or to Victorian times. The urgency to remember and record the past, before the bearers of these memories die, is striking. So many of 'the old ways' of doing things, and of the major industries, have vanished from the Firth, and so many younger people have moved away to find further education or employment: does this add disproportionate weight to the looking back?

*

We are being reminded daily of the damage we are doing to our planet: causing the extinction and near-extinction of animal and plant species across the world, on land and in the rivers and seas – either by our direct action or indirectly, through the widespread effects of the rise in temperatures and the changing patterns and extremes of weather. On a personal note, sometimes the melancholy and sense of helplessness can be almost overwhelming, especially while watching the numbers and species of fish and marine invertebrates plummet around our coasts, and witnessing the seeming indifference to the existence of other living creatures around us – both in the rural area where I live, and along the Solway. There's even a word for this particular type of grief and melancholy, solastalgia (from 'solas' at the root of 'solace' and 'algia' or pain, with hints of 'nostalgia'). We're trashing the neighbourhood, not just for the long-term residents, but for the visitors too.

on a Solway flat

winter migrants gather
in long black lines
along a silver sleek

heads held back,
throats
thrust toward
an onshore rush[7]

Wildfowling – the killing of visiting migrant geese, ducks and waders, by shooters on foot or in punts – has long been part of the Solway's heritage along the marshes and merses of both shores.

One dawn back in January 2011 I went out onto a marsh with wildfowler Brian Hodgson, then the secretary of the South Solway Wildfowlers Association. My two alarm clocks wake me at 4.45 a.m. and I pile on thermals and multiple layers of warm clothes. I'd previously spent hours reading Eric Begbie's reminiscences of his wildfowling days on the Firth of Forth, especially his chapter 'Be prepared', and another that might have been titled 'How to survive hypothermia and total immersion'.[8] I throw a bag of dry clothes in the car and, bearing in mind also his comment that 'the white orb of a human physiog stands out like a glowing beacon amid the drab colours of the marsh', pull on a dark, hooded fleece. There has been a heavy frost overnight and the roads are sparkling white in the light of the nearly full moon; black tyre-tracks show that I am certainly not the first to travel those roads this morning. It is cheering to discover how many other people are up and about at that early hour – there are lights in bedroom and bathroom windows, and yard lights at farms. A shooting star arcs across the sky and I make a wish then briefly stop the car on a long straight road and turn off the lights, to enjoy the emptiness. The landscape here is as wide and flat as the East Anglian fens.

Brian is already waiting at our arranged meeting point in the pub car park at Abbeytown and we drive along a single-track road to a dead-end near a farm on Border Marsh. We are not the first here, either; a black pick-up is parked in the lay-by, its owner, like us, preparing to set off across the marsh. We are accompanied by Lucy, a bouncy and sociable black labrador who is clearly delighted to be out on the marshes. The grass is white and crusty with frost, and the moonlight makes it easy for us to pick our way across frozen pools and ditches, heading in the direction of the Anthorn masts. We eventually reach a series of branching creeks known as Pierrot's Hole and we settle down, our boots making sucking, popping sounds when we shift our feet. Brian spreads out a bed of camouflage net for Lucy and she curls up, pressed against the muddy wall of the creek, and is soon snoring softly. We lean back against the wall, chatting quietly, our elbows propped on the frosty turf, and wait. The moon sinks behind a low wall of cloud to the right of Criffel across the Firth; the mist that has been hiding the lower lights of the masts rises so that the upper lights are haloed in orange. The silence is complete. The mud is icy cold.

'There were ice floes piled in the creeks around Christmas, some of them were the size of snooker-tables,' Brian says. 'You should've heard the noise they made when the tide was coming in.' There were ice floes drifting with the tides and piled on the shore of the Firth, too, and the spectacle even hit the national news.[9] That long period of extremely cold weather, when the fields and marshes and fells were coated with snow and ice, meant that the geese had great difficulty grazing and were becoming hungry and thin. Geese migrated to, and stayed on, the Solway in great numbers, but the South Solway Wildfowlers Association (WA) asked its members to observe a limit on the number of geese that could be shot, no more than three geese in any twenty-four-hour period. That limit no longer applies on this January morning, but at present there are no geese near us.

'*Phweet*', the characteristic whistle of the male wigeon; curlews, an occasional lapwing; the bubbling twitter of sandpipers; the raucous '*quack-quack*' of a mallard . . . and then we hear a conversation starting, a gabbling, clearly a large flock of geese across the water at Cardurnock Point. Barnacle geese. Now the gulls join in, and there is a harsh croak of a heron. But the most exciting sound is the '*wink-wink*' of the pink-footed geese. Out there in the darkness the wildfowl are talking, amongst the silt and pale grass of the saltmarsh, and on the water; there is a sudden breathless flurry of wings overhead as the ducks return to the Solway. Somewhere out there are hundreds of birds getting ready for another day of feeding; for the geese, another day of preparing themselves for that long migration northwards again to breed. Brian and I are talking quietly, leaning back against the bank, when he suddenly leaps forward, grabs the shotgun and – before I have realised that about twenty wigeon are flighting over – he fires off a single shot. The formation wavers and the birds jink to one side then carry on, but none falls. Lucy looks at me reproachfully.

As the sky turns from black to grey the dark marsh to the left has taken on a steely shine; the main creek is filling up and spilling over into our side-branch, an insidious silent creeping of brown water, débris swirling gently on the surface, pale bubbles of froth. The water reaches our feet but not Lucy's backside, and washes gently back and forth, then after about thirty minutes it silently drains away. We are waiting for the geese to leave their roosts and fly inland to feed; at dusk they will fly back to the coast again. Duck do the converse – they feed inland at night and return to the coast at dawn. Each day, then, there are two 'flights' when wildfowl may be shot.

I had had an unpleasant mental image of a man with a shotgun standing triumphantly with his foot on a pile of feathered corpses. I certainly hadn't known what to expect and was relieved that this would not be the case, because only certain species of duck and goose are legal quarry and there are limits on the numbers that may be taken: you may not, of course, shoot at curlew or sandpiper. Greylag, pinkfooted and Canada geese are quarry. The barnacle goose is definitely not – the whole precious population of barnacles from Svalbard over-winters here. Mallard, wigeon, teal and pintail, are permitted quarry, but shelduck, eider and scaup are protected. But how can you tell the difference when it's dark? 'The general rule is, if you can't identify it, don't shoot it,' Brian says. 'But you can usually tell because you'll have heard the calls.' He has been wildfowling for nearly thirty years and when I suggest that he must have learnt a lot about the behaviour of the birds, he agrees but points out that, 'It's always different, it's influenced by so many things – the phases of the moon, the height of the tides, the wind. If the wind had been driving the tide high onto the marsh this morning, the duck would have come right up to the edge here instead of staying out on the water.' (We can see their dark shapes bobbing further out.) 'Or when there's a full moon and it's a clear night, the duck will be feeding on the flashes [the shallow ponds on the marsh] – we'll put out decoys, and we'll be out all night ourselves. But you need a bit of cloud for when the geese flight, you need to be able to see their silhouettes against the cloud.' We hear a couple of pinkfooted geese calling, and one even flies close by: but not close enough. Brian tries to encourage it to approach by blowing one of the two whistles that hang around his neck. These 'calls' are for wigeon, and for pinkfeet and greylag, and sound astonishingly realistic, but the goose is not interested. Now the sky near Skiddaw is pale apricot, and I can see the sea, no more than twenty metres away. The geese across the water have been silent for a while and I have given up on seeing them but suddenly they all begin calling again, and then we see the whole flock flight. The sight and sound of them, the massed dark shapes lifting into the air, is thrilling – and they fly away from us, towards the RSPB's Campfield Reserve. There will be no more shooting here today but Brian doesn't seem to mind, because I'm beginning to learn that for him and many other wildfowlers the shooting is a bonus, not the real reason for coming.

'I just like being out here and being part of it,' he says. 'The early morning, the solitude of it. It's an environment that's alien to us, and ninety-five per cent of the time the wildfowl have the advantage. There was one morning when about four thousand pinkfeet lifted off. "What a sight!" I said to the dog. "Look at that, and if you only remember one thing in your life, remember that!"'

We have heard curlew, wigeon, redshank and shelduck, and flocks of gulls are now beating their way up the Firth; we have experienced the wildness and uniqueness of the marsh at dawn. And then, as we return to the cars, a lone pinkfooted goose flies towards us. Brian calls it with his 'goose' whistle, and it comes closer – overhead, and within a stone's throw. It takes a good look at us then veers off. The goose, like us, is probably laughing. Brian is apologetic that I haven't been able to see flocks of geese flighting over our heads. 'The moon's upset things, it all gets back-to-front when there's a full moon because the geese go off and feed inland during the night. But it's coming up to the end of the season and the moon will be right, so we should get a few good mornings. There might be a dozen to fifteen people out then.' The owner of the black pick-up returns as we're pulling off our water-proofs. He's had no luck either but nevertheless he's smiling broadly. 'A long walk for nowt,' he says. 'There must've been four to five hundred geese over there, but they weren't going anywhere.'

We drink our very welcome hot coffee and Brian says that the lack of birds today might be more to do with the clear, dry dawn – stormy weather would drive the birds off the shore and force them to fly lower, giving the wildfowlers a rare advantage. Although when he describes it thus, 'When you hear the rain and wind battering against the window at four in the morning, you know you've got to get yourself out of bed and get out there on the marsh!', I'm not sure whether hunter or quarry would be most disadvantaged. To go wildfowling you must have a registered shotgun and be insured and – here on the southern, English, shores of the Solway – have a permit from the South Solway Wildfowlers Association. The organisation is affiliated to BASC (the British Association for Shooting and Conservation), and oversees more than 8,000 hectares on the Solway marshes stretching from Grune Point at Skinburness, through Calvo and Border, Newton Arlosh, Cardurnock, Drumburgh and Burgh, with a lease on part of Rockcliffe marsh. About three-quarters of this is in the form of reserves, like Raby Cote marsh, which ensures that there are large areas where the birds can roost and feed undisturbed.[10]

I first heard about the Solway's wildfowlers from a conservationist who was also a shooter, and was puzzled by this schizophrenic mix of 'poacher and gamekeeper'. The Wildfowl and Wetland Trust was set up by Sir Peter Scott, himself a wildfowler, and BASC's name also links shooting with conservation, purportedly in the context of culling sick animals or keeping population size within sustainable limits. So how could this possibly work with wildfowling? The South Solway WA, for example, works closely both with Natural England – digging scrapes or flashes and cleaning existing ponds – and with the Marsh committees who look after the grazing on the stints. As always, there is a conflict between the needs of the graziers, who need good drainage to improve the grass, and the wildfowlers and conservationists, who want to keep the marshes wet; as always, there must be compromise.The wildfowlers also have a wardening team, who keep an eye out for poachers and maintain the gates and car parks at the access points to the Marsh. Curiously, the ducks but not the geese can be sold on; geese must be eaten by the fowler and his friends. This reminds me of a book I saw advertised on the Internet: *How to carve wildfowl*. I had expected a cover photo of a glistening, well-basted goose breast, a still life with carving knife and fork and a jug of gravy. But it was about wood-carving – making decoy birds. Decoys generally resemble mallard and are set out in a row on a tethered nylon thread, to bob about with the tide and convince incoming flights that all is well and this is a good place to visit.

Wildfowling had always been more of a working-man's sport, Brian Hodgson had told me, and during hard times had provided many families with food. But there is no denying that wildfowling *is* a sport. Yet, as so often with those who 'use' the Solway, it seems to be the wildness and emptiness and wildlife that are important to the wildfowlers. A Cumbrian wildfowler recently told me, 'The main draw is just to be out on the marsh – to hear the birds awakening and see the sun rise,' and David Campbell, former chairman of the Scottish Solway Wildfowlers Association, agreed: 'Shooting geese isn't the main reason they do it, it's not the be-all and end-all. It's the being there, in a wild place – it gets in your blood. You either love it or hate it!'

And for Brian: 'There's been a couple of occasions, when the dog's brought back a goose and I've laid it down here on the bank, and I've looked at it – and for a split second, I've wondered why I'm doing it. Just for a split second, mind. And afterwards I've thought I must be getting soft, but when I've mentioned it to some of the other lads, there's one or two of them have thought the same. Just that quick thought . . .'

Wondering whether attitudes amongst wildfowlers have changed in the past decade since I went out on the marsh with Brian, I recently revisited Caerlaverock. There is a major difference between wildfowling in England and Scotland – in England, wildfowling requires membership of a syndicate or club, but in Scotland anyone with a shotgun certificate can shoot wildfowl on the foreshore, the area between the low-water and high-water marks of an ordinary spring tide, unless in areas covered by a bye-law. Of course, this doesn't mean shooters with shotguns are prowling everywhere around the Scottish coasts: it applies to the estuaries and coastal marshes where ducks and geese are roosting and feeding – Forvie, Montrose and Findhorn, for example, and along the Solway at Wigtown, Mersehead, Auchincairn, Annan: and at SNH's Caerlaverock Reserve.[11]

Here, again, is the conflict – between the reserve's long-term purpose of protecting birds and other living creatures, and the continuing need to accommodate wildfowlers. The over-wintering birds like the pinkfooted and barnacle geese often rest out on the water over Blackshaw Bank during the night. As someone told me, on a 'calm tide, they float about with their heads tucked in' and someone else pointed out, 'They're well-insulated and they feel safe from predators out there.' Caerlaverock has bye-laws that protect the mudflats – shooting is only allowed over a fairly small area of the reserve's foreshore. For many years the Caerlaverock permit scheme has been a rôle model for Scottish wildfowling, in that shooting is highly regulated: local people and visitors must apply for the limited number of permits; only a fixed number of shooters are allowed for each flight of birds; there's a fixed 'bag' limit of five birds per flight; records must be kept and returned; and shooters will be refused future permits if the reserve's and BASC's rules are not adhered to; the police may even be informed of breaches. There are volunteers, themselves wildfowlers, who monitor the behaviour of permit holders for most of the flights during the season. It is well-managed, and feedback from permit holders shows appreciation for respectful and good behaviour at Caerlaverock, I was told, but maintaining these standards requires a large amount of time from the monitors and the reserve officers, a resource that is already stretched, to meet all the demands of reserve management.

One late-autumn afternoon when I was driving up the Cumbrian coast road, there was a strangely elongated dark cloud far ahead that was moving very swiftly across the sky, heading across the Solway. A friend in Dumfries had seen it too – an enormous flock of geese. Perhaps it was then, in late October, that the watchers at Caerlaverock reported the first large flock of 3,100 pinkfooted geese had arrived. 'Five birds bag limit': again the image of piles of feathered bodies came to mind. According to the SNH Caerlaverock Reserve report for the 2018/19 shooting season, no mallard, shoveler or gadwall were shot, but one greylag and three wigeon were killed – and 221 pinkfooted geese.[12] To the east of the reserve, the Scottish Solway Wildfowlers' Association manages the shooting over Priestside and Brow Well by Lochar Water; permits and bag limits are required here too. But further east again, there are no restrictions and no permits are necessary – the normal foreshore shooting permitted in Scotland can take place. And this is where the present differs markedly from the past: the increasing rôle of social media. I've been told that, 'The word goes out, on Facebook or wherever – "Saw lots of birds at X yesterday" – and mobility is so much greater now, people are prepared to drive long distances. These days you can easily access inexpensive firearms and you don't need training in how to use a shotgun.' David Campbell explained that, on the Scottish Solway wildfowling areas like Priestside, 'Someone may be out and get a good flight in the morning, and by the afternoon I'd have people turning up at the house to get visitors' permits – but we've changed that, they have to apply by post now. But on the shore where there are no bye-laws, you can get a lots of guns at times. They misbehave, there's high shooting – there can be quite a lot of people shooting and it sounds really bad.' Even where shooting is controlled, as on the Cumbrian side, Brian Hodgson told me that the South Solway Wildfowlers don't allow anyone to post about the number of birds on Facebook.

So are present-day attitudes to wildfowling different from a decade or so ago? In Scotland yes, through the effect of social media and mobility. It's notable that both wildfowling and 'twitching' use the same tools of communication and rapid transport – twitchers descend on Shetland as rapidly as peregrines when some unusual bird has strayed from its normal migration route. Additionally, this was the first time during my twenty years or so of wandering around the Solway that people were sometimes reluctant to talk to me or to have remarks attributed to them. There is, then, the 'romantic' idea of wildfowling – perhaps typified by my earlier outing with Brian Hodgson – as against the present-day reality. 'The traditional view, of the old-fashioned wildfowler, is that you learn your fieldcraft from Dad, who teaches you about the birds and the environment, and you move up through the ranks,' someone explained to me. 'And against that now you've got the other people, who want to shoot as many birds as they can. It's like a computer game, it's like target practice. Wildfowling is a sport, yes, but it's how you behave in that sport that is important.' Another wildfowler (from the English side, and also someone who has a job in conservation) told me that there were 'a few cowboys out there [on the Scottish side] – it's not the best of situations'. And 'There are wildfowlers – and there are shooters.'

'Poor behaviour' is a phrase that now comes up frequently, not just in conversation, but in reports in the press too: not just on unregulated parts of the Scottish Solway coast but also further to the north-east, in places like Findhorn. Some people would like to see the Scottish law about access for shooting changed. There is, too, among local wildfowlers the 'sense of belonging to a space', as a community. Visiting wildfowlers, especially the poorly behaved ones, can be seen as the enemy, who are giving the sport a bad name.

Perceptions of how we should treat other living creatures have, rightly, changed. There is a resurgence of the ideas of animism, a growing recognition that wild animals have rights. 'People say, "The geese have travelled all this way – what right do we have to shoot them?",' I am told. And for Barry Lopez, 'Birds tug at the mind and the heart with a strange intensity. Their ability to flock elegantly as the snow goose does, where individual birds turn into something larger, and their ability to navigate over great stretches for what is for us featureless space, are mysterious, sophisticated skills.'[13]

Stretching our gaze north, far over the horizon to the Arctic, we should also worry about how the breeding and nesting grounds of these geese will be altered as the tundra warms and its vegetation and insect populations change, and the web of interactions is pulled apart and broken. Here to the south, as the ice sheets melt and the sea level rises relative to the land, there will be changes in the saltmarshes and low-lying land where the geese come to spend the winter. These are rapid changes that are happening in our lifetime, and we should worry that the geese and other migrants will not have enough time to adapt.

*

The Solway's merses and mudflats are being continually altered by the tides. 'Chaotic and unpredictable', 'One of the most aggressive estuaries in the UK' and a 'Highly dynamic estuarine environment': the Solway has a strong personality (see Introduction). The UK must deliver 15 per cent of its energy from renewable sources by 2020, under the European Union's 2009 Renewable Energy Directive, and the sea is the ultimate 'renewable' source of power. Our island is surrounded by an unlimited store of energy carried by the tides and the waves, energy that can be captured by turbines and converted into electricity, and the Solway Firth is potentially a magnificent source of this energy. Proposals for harnessing its tides stress its large tidal range. The difference in height between high and low tides on the Solway can be as much as ten metres or as small as about five metres depending on whether the moon and sun are in alignment (spring tides) or at right-angles (neap tides). Sometimes the spring tides are especially large, as during the 'supermoon' phase at the end of September. During the flow of these spring or 'big tides' as they're known locally, a massive volume of water has to pass up the Firth (and out again) during each approximately twelve-hour cycle; the rate of flow is impressive and daunting to watch if you are on the lower shore, and the bore in the

Upper Solway near Bowness can be seen and heard, and even experienced, as I found when I went haaf-netting. It is not a large bore, but perhaps was greater in former times, for Neilson (1899) describes how 'the sea approaches [the River Eden] with great speed, gaining as it goes; the wave is white with tumbling foam; a great curve of broken surf follows in its wake; and the white horses of the Solway ride in to the end of their long gallop from the Irish Sea with a deep and angry roar'.[14]

Earlier, according to the compilers of the 1848 *Gazetteer*, the incoming tide and its bore were even more dramatic: whether this is a true account of the phenomenon or the author's delight in his powers of imagination can only be guessed at:

All its tides are rapid, and constitute rather a rush or careering race than a flow or a current of water. A spring tide, but especially a tide which runs before a stiff breeze from the south or south west, careers along at the rate of from 8 to 10 miles an hour; it is heard by people along the shore upwards of 20 miles [*sic*] before it reaches them, and approaches with a hoarse and loud roar, and with a brilliance of phenomena and demonstration, incomparably more sublime than if the wide sandy water were densely scoured with the fleetest and the most gorgeously appointed invading army of horsemen; before the first wave can be descried from the shore, a long cloud or bank of spray is seen, as if whirling on an axis, and evanescently zoned and gemmed with mimic rainbows, and the rich tintings of partial refraction, sweeping onwards with the speed of a strong and steady breeze.[15]

Brian Blake is wryly prosaic: 'The notices in certain places around the Solway do not exaggerate when they warn bathers (and naturalists) of swift and dangerous currents.'[16] On the marshland road near Port Carlisle, notices warn too of the sudden spilling-over of the rising water – signs, their black letters wearing away (or perhaps picked off by someone who had a penchant for collecting Hs), note that W EN

WATER REAC ES T IS POINT MAXIMU DEPTH IS 1 FOOT (it may even be two). When the tide reaches the innermost reaches of the Firth, the difference in height between high and low water is much reduced as the water meets the river flow and spreads out over the salt-marshes and mudflats; here at low water in the big spring tides most of the water flowing is from the rivers Eden and Esk. So what is the future for harnessing the energy of the Solway's tides?

The basic principle of harnessing tidal power is that a proportion of each incoming and outgoing tide is forced to pass across the vanes of a turbine, causing the turbine to rotate against a magnetic field within a generator and produce electricity. The turbine must be bi-directional – it must be able to work with both the ebbing and the flowing (incoming) tide. It must also be unaffected by high densities of suspended sediment. Power generation will only occur when water is moving at a prescribed range of flow rates through the turbines – but tidal flow rates are not constant throughout each cycle, there are the periods of 'slack tide' each side of the time when the tide turns, so the period of generation is between six and ten hours, depending on the location. This can be improved by trapping some of the water and releasing it, so that the available energy depends on the head of the water above a turbine and the volume of water flowing through it. If the tidal flow is constrained by walls such as a barrage or a lagoon, water levels can be controlled by gates and overflows to provide a manageable flow rate through the turbines.

With a lagoon, some of the water is trapped within an enclosed space while the rest of the main tidal flow passes by outside; with a barrage across an estuary, the tide must pass via the turbine spaces and also, presumably, gates built into the wall. So what does building a tidal power-generating system require, with reference to the Solway? Turbines, obviously – which need to be set in a supporting framework, which might be a rigid wall; a generating system; transformers – which might be installed nearby (as at Robin Rigg offshore wind farm); cables to transport to the shore the electricity generated; onshore control centres and support systems; links to the National Grid that have the capacity to carry the electricity generated (and that capacity is only on the English side). Walls – barrages, lagoons – need foundations on the seabed so, since large areas of the Solway's bed are very labile and changeable, bathymetric and geological surveys are obviously necessary. Walls need cement and boulders or 'geo-tubes' filled with rock, and baffles to dissipate the power of the sea during storms. For rocks and boulders you need a very large source of stone (perhaps from a former granite quarry on the Galloway coast). Also vital is the infrastructure for transport, via the sea (such as the port of Workington) and road and rail. There is the passage of local shipping to consider too – ranging from tankers and other merchant vessels visiting the ports of Workington and Silloth, to trawlers, scallop boats, wind farm support vessels and pleasure yachts. Another requisite which is rarely mentioned is an understanding and modelling of the potential effects of altering the currents, and of the building process itself, on the erosion and deposition of sediment. And the final requisite is serious, dispassionate thought about what is valuable about the Solway Firth and what we, as the human species, 'want' from it.

Here, then, is the conundrum: we know we need to produce energy by renewable means, to escape our dependence on fossil fuels; we are going to need to generate much greater quantities of electricity in the future – think of all those electric cars, and the gigantic amounts of power that the banks of servers associated with the Internet, AI, 'smart homes' and data-mining will consume. Tidal power schemes are a good way of doing this; and with regard to the local economy on both sides of the Firth, major and long-term construction schemes bring in money and much-needed jobs. On the other hand, such major interventions are potentially very damaging to the Firth and its smaller estuaries, especially the margins, and to the resident and visiting wildlife – and swathes of the water and the margins are protected.

First, a quick trip around the greater Firth's perimeter, from Irish Sea to Irish Sea (this can be done online using MagicMap[17]), shows large areas – bays, marshes, mudflats and seabed – where the environment is protected by international, European, national and local designations; some of these 'protections' have legislative teeth, others are all too open to abuse.[18] Many of the larger areas are designated because of their importance for food and shelter for wading birds and winter migrants like whooper swans and the pinkfooted and barnacle geese; the resident invertebrate animals are simultaneously protected. Other designations might specifically care for creatures like the honeycomb worms or migrating sparling.

The tidal power schemes proposed for the Solway are relevant for any Firth or major estuary. 'Barrage' is a dirty word with regard to generating tidal power. For a Firth that is 'aggressive', 'highly dynamic' and laden with sediment, putting a great wall across its mouth will cause major changes in hydrography, tidal flows and sediment deposition. It is considered an outdated and disruptive technology, very expensive to construct, with too many problems relating to environmental change and remediation. North West Energy Squared (NWE2) proposed to construct barrages (or, as they confusingly called them, 'gateways') across several west-coast estuaries, including the Solway from Workington to Kirkcudbright, back in 2015, but the idea seems to have been quietly dropped.

For the past few years I have been following the developments (or lack of them) through reports and conversations with the people proposing different projects for the Solway; the details are interesting but not necessary to repeat here, other than in outline. There are the lagoons – Russell lagoons are U-shaped, each end joined to the coast; Ullman lagoons are 'doughnuts', built out in the estuary with no connection to the land. As far back as 2011 Tidal Lagoon Power[19] started working on a proposal for a Russell Lagoon at Swansea with the idea that, if this proved successful, they would extend their system to five other estuaries, including the Solway Firth – they would construct a 31-kilometre lagoon wall from Workington as far east as Dubmill Point on Allonby Bay. Charles Hendry's review in 2016 of tidal power suggested strongly that Swansea should be a 'pathfinder' project – in other words, build it, let it run for five years, see how the finances work out and (importantly) measure its effects on environment.[20] Since then the process has been tortuous, with major steps backwards and minor steps forward, and the latest amusing incarnation of the Swansea scheme is the Dragon Energy Island, along the lines of Dubai's Palm Island. Should the Swansea project ever go ahead, and should it prove its value as a pathfinder, Cardiff would likely be the company's next objective: the Solway would probably drop off the list. Another Russell lagoon plan, an 18-kilometre wall from north of Maryport to Mawbray, and around the outside of the Allonby Bay Marine Conservation Zone, has been suggested by the engineering and construction firm Arup.

Both lagoons would protect the coast and its road from further erosion. But in both cases, financial backing from the Westminster and devolved governments is seen as important in attracting further investment and getting the practicalities underway. Meanwhile, the Tidal Electric Consortium, backed by American hedge-funders, are proposing one or two Ullman lagoons out in the Firth – one on the Galloway side near Heston Island, another near Allonby Bay.[21] There is also Solway Energy Gateway's proposal for an 'electric bridge'[22] that would involve weir-like plates that could be raised or lowered to retain the tide, with a new type of turbine based on the differences in water pressure created by the Venturi effect. This scheme would be much higher up the Solway, along the line of the former viaduct from Bowness across to Annan (where it is possible to walk across at low tide); the proposer is keen that an actual bridge across the Firth for cyclists and pedestrians should be incorporated too. Between the closing of the viaduct in 1921 and its dismantling in 1934, it remained a very useful route between the two countries:

TRESPASSING ON SOLWAY VIADUCT

Prohibition in the town of Annan on Sundays had a sequel at Carlisle Police Court on Saturday, when twenty Scotsmen were fined for trespassing on the Solway viaduct. It was stated that every Sunday night Scotsmen made a practice of crossing the Solway Viaduct from Annan to Bowness-on-Solway to obtain drink. Much damage had been done to the permanent way, and there is grave danger of trespassers falling through gaps in the bridge into the sea below.[23]

Newer licensing laws mean that Scottish pubs are no longer prohibited from selling alcohol on Sundays!

As for timescales: even if a lagoon, of either sort, were given approval tomorrow, the time to power generation is in the region of about seven to eight years. Of this, the scoping, modelling, working out risk mitigation, gaining the necessary permissions, public consultations, planning applications, agreeing Contract for Difference price (the unit cost of the electricity produced) with the government and so on would take three to four years. The actual construction phase – requiring movements of large amounts of material by land, rail and sea; dredging and moving sand; cable-laying; building the walls and turbine housing; building onshore works and offices; environmental 'remediation' and so on – would take a further three to four years.

So, perhaps by about 2030, the characterful Solway will have been persuaded to lend some of its power for the generation of electricity. In the meantime, new technologies are being developed elsewhere, particularly in the Pentland Firth – turbines anchored to the seabed by tripods, or floating on the surface to harness the tidal stream: generating systems that might not be so efficient but are likely to be much less damaging to estuaries and their edges. And in the meantime, too, the sea level will be rising and extreme weather events, including storm surges, will be a much more common occurrence. Why have we – on an archipelago of islands, surrounded by the waves and tides – left harnessing this renewable energy too late?

There are, then, many known unknowns (Donald Rumsfeld's taxonomy of the unknowns should not be sneered at) with regard to the Solway's future. It *is* known that '[The Solway] is very, very unpredictable, it's uncontrollable' (see Introduction) and that human interventions have frequently had unexpected consequences. There is a high probability that the disruptions caused by lengthy construction work will have damaging effects on marine and intertidal life, and the lagoons will affect the tidal flows and sedimentation patterns. One projection of sea-level rise for 2050 shows that much of the Upper Solway's Cumbrian coast and the regions along the Nith and at

Caerlaverock will be at risk from inundation.[24] Sea-level rise relative to the land will be less here than in the south of England, because Scotland will continue to rise – but nevertheless the rise will be significant. The Solway will be wider, more of its liminal areas will be underwater at high tide. The tides, of course, will remain the same, but the extent of wave damage will possibly increase too. This will need to be taken into consideration when designing a tidal power scheme.

Additionally, rising sea level will affect the outflows of the rivers. The profiles and positions of saltmarshes and mudflats will alter. Changes in relative sea level and the labile margins have always been part of the Solway Firth's stories (Chapter 2), and there has to be hope that the numbers and populations of the non-human residents and visitors will adapt. But with the numbers and species of waders that visit the protected areas already decreasing, and the numbers of salmon and sparling in decline . . . as a friend of mine who works for Cumbria's Wildlife Trust said, 'This might just be sufficient to tip some of the species over the edge in the future – they won't come back.'

*

There are deep pools in the wave-rippled sandstone at Fleswick Bay, pools which have been excavated by pebbles that have been rotated year after year by the waves. The smooth walls are speckled with barnacles and the pink, rigid tufts of *Corallina*, small coloured stones shine at the bottom, and on a windless day the surface of the water is invisible; the hollow is dark and still and, like Peter in Neil Gunn's *The Well at the World's End*, you might be persuaded that the well is dry:

> With the odd feeling of having hunted the wrong well in another world, he turned abruptly away and went striding down the same path looking for the right well . . . For a man can always see where the surface of the water, however crystal, touches the sides of a container. There *is* a difference between air and water. All he had got to do was pick up a handful of dry pebbles from the

bottom. But he could not move and his eyes stared as though fascinated by some invisible spirit in the well. Then with an air of one on the brink of some extraordinary revelation he stooped and slowly put his hand down, and his hand went into water.[25]

There are such deceptions here, too, at Dubmill Point, where the *Sabellaria* reefs have trapped the outgoing tide. I have halted at the edge of silent pools in the still of the early morning and watched the rays of the sun lighting the top of Criffel across the Firth, and then sweeping down across the water, enfolding the shore and warming the dark shapes and the silvered pools of the reef. Colourful lives are suddenly revealed. At such times the vastness and microscopic details of this seemingly 'other world' are almost overwhelming.

Sometimes I need to escape this tension to stride along the midshore, focusing on the distance – the macro-details of rolling sandwaves and the sliced-open faces of dunes. But on this morning I'm halted by a figure in the distance, down amongst the reefs at the very bottom of the shore. She – I am sure it is a woman – is moving slowly, bending down to look into pools and to touch and pick up objects. In all these years I have never seen anyone else looking so closely at the *Sabellaria* reefs and pools. But I'm unnerved, too, because she has white hair, she is wearing a jacket that is an identical plum-red colour to one I used to have, and she is accompanied closely by a black-and-white border collie. It is as though I am seeing my doppelgänger.

I carry on walking but after about ten minutes I turn back, and I see that our paths will now intercept – the dog runs towards me and politely allows me to stroke its head. I see, too, that although the woman has a slight limp she is only a little older than I am and, because dogs (and small children) are always ice-breakers for conversations, we start to talk about the peace . . . the delicate sandy worm-tubes on the reefs . . . she has found coatings of green sponges and seen the strange shapes of burrowing anemones . . . she shows me a mussel shell with a tiny hole with bevelled edges, bored and eaten by a dog-whelk, and asks if I'd like to have it . . . I thank her, but have a dozen or so already . . . Her white hair is wiry and her face thin and weather-beaten and although we smile a lot at each other, her eyes are fierce. Our lengthening conversation (the dog mutters softly, and sits) shifts to the councils' wishes, each side of the Solway, to increase tourist numbers.

'I don't want anyone else to come here,' she says, 'I like the beach when it's empty. And what would they *do*, anyway?'

And because I have been looking out across the water and trying to imagine a lagoon wall, I tell her about the tidal power proposals, and how we need more electricity to be generated from renewables, but . . .

'But the planet's *fucked*!' she says, frowning, looking at me quickly to see if I'm shocked. 'It's fucked. It's all too late. And anyway there are far too many of us – the best thing would be if our whole species was wiped out, then the planet could recover.'

Soon afterwards she hirples off along the shore, and I wish we could meet again in the future. I wish that I'd asked her about herself, and I wish the polite dog were mine so I could teach it to 'come by!'

I keep thinking, over and over, of Simon Armitage's thought-child, who said

If I breathed the word

that disappeared all people
in the world,
leaving the world

to the world, would you
say it? Would you
sing it out loud?[26]

'Leaving the world to the world.' The Solway's character will change again; it will always enjoy its unpredictability. And there will continue to be mudflats along the Upper Solway – and the mudshrimps, *Corophium volutator*, will continue to thrive.

References

General books and reports about the Solway include:

Brian Blake, *The Solway Firth* (London: Robert Hale Ltd, 1955).

A.J. Scott, *Solway Country: Land, Life and Livelihood in the Western Border Region of England and Scotland* (Cambridge Scholars Publishing, 2015).

The Solway Firth Review, produced by the Solway Firth Partnership (1966). A digitised version is available online at www.solwayfirthpartnership.co.uk/about-us/what-we-do/, and a new edition is in production.

A.A. McMillan, J.W. Merritt, C.A. Auton and N.R. Golledge (eds), 'The Quaternary Geology of the Solway', *British Geological Survey: Geology and Landscape (Scotland) Programme Research Report* Rr/11/04 (2011). Available to download via NERC Open Research Archive: http://nora.nerc.ac.uk.

NE 536. National Character Area No. 6, The Solway Basin, Natural England (2015). Download from http://publications.naturalengland.org.uk/publication/5276440824119296.

Notes

EPIGRAPH

1. From Norman Nicholson's 'Five Rivers' (1944) in *Collected Poems*, ed. Neil Curry (Faber & Faber, 1994).

INTRODUCTION

1. Andrew Lysser's company Cumbria Gyroplanes was based at the Carlisle Lake District Airport.
2. On the Solway, the big spring tides usually occur at the unsociable hours of 6–8 a.m. (and about 12 hours later). The reasons are complicated. When the sun and moon are in alignment (syzygy) – either at the same side (conjunction) of the Earth or at opposite sides (opposition) – the tidal bulge is greatest, giving us the spring tides. When the moon and sun are at right-angles (quadrature), the tidal bulge is smaller, giving us neap tides. The Earth rotates on its axis every 24 hours. The moon revolves around the Earth–moon centre of mass every 27.3 days. Therefore the period of the Earth's rotation with respect to the moon – one lunar day – is 24 hours 50 minutes. This is why the times of high water advance by about 50 minutes each day: the interval between the two high tides each day is 12 hours 25 minutes; the interval between high and low water is 6 hours 12.5 minutes. But the observed tide is the sum of a number of 'harmonic constituents' or partial tides, 'each of whose period precisely corresponds with the period of some component of the relative astronomical motions between Earth, Sun and Moon' (from *Waves, Tides and Shallow-water Processes*, Open University, 2008). There are up to 390 harmonic constituents, ranging through semi

-diurnal, diurnal and 'longer period'. Here in Britain our tides are on the 'semi-diurnal' pattern, where the tidal range fluctuates unevenly throughout the month. So, to think about spring and neap tides we only need to consider two of the semi-diurnal constituents: principal solar, S2, with a period of 12 hours, and principal lunar, M2, with a period of 12.42 solar hours. These are the most important ones because they control the spring–neap cycle. This means that the highest and lowest spring tides will occur at the same times of day for a particular location because the period of the S2 constituent is 24 hours. For more on tides, see the EasyTide website, and Hugh Aldersey-Willam's book *Tide* (Penguin, 2016).

3. The Google satellite view stitches together images from satellites such as LandSat, NOAA, the US Navy satellites and the Sentinel/Copernicus series. The information can be found in a small bar on the bottom right-hand side of the image.

4. *Crossing the Moss* (www.crossingthemoss.wordpress.com): the story of the Solway Junction Railway, the Solway viaduct, and the peatlands ('Moss') of Bowness Common. A project that I and my collaborator James Smith carried out in 2016/17, funded by Solway Wetlands Landscape Partnership and the Heritage Lottery Fund.

5. Moricambe Bay, north-east of Silloth on the Cumbrian side of the Upper Solway, should not be confused with Morecambe Bay, which is to the south of the Lake District, between Lancaster and Morecambe.

6. See Chapter 9 for more about the Lockheed Hudsons and Moricambe Bay.

7. For more about RAF Dumfries see the website and booklets for the Solway Military Trial: www.solwaymilitarytrail.co.uk.

8. Kathleen Jamie, 'Basking Shark', from *The Tree House* (Picador, 2004).

9. All life-stages of natterjack toads, *Epidalia calamita*, have full legal protection under the UK Wildlife and Countryside Act, making it an offence to kill, injure, capture, disturb or sell them, or to damage or destroy their habitats.

10. There are volunteer organisations that work to clean up such rubbish. SCRAPbook is a project of the Moray Firth Partnership assisted by the Solway Firth Partnership (see map of the SCRAPbook activities at https://marine.gov.scot/maps/1854); and the UK Civil Air Support (SkyWatch), https://civilairsupport.com/, carries out aerial searches and photography and other observations and recordings. One of the flights and beach clean-ups was on BBC4's *Our Coast*, episode 4, Dumfries and Galloway, https://www.bbc.co.uk/programmes/m000frj7.

The Solway Firth Partnership (SFP, www.solwayfirthpartnership.co.uk) is an independent charitable organisation which works cross-border, covering an area of sea and coast from Stranraer on Loch Ryan down the Scottish coast and across to St Bees Head. Based in Dumfries, the SFP aims to bring people together across and around the Firth to achieve sustainable development and environmental protection, and to support integrated marine and coastal planning and management.

11. The (English) Sea Fisheries Committees became the Inshore Fisheries and Conservation Authorities (IFCAs) in 2011. The fisheries' bye-laws for the English side of the Solway (covering the area up to six nautical miles from the coast) are overseen by the north-west IFCA. (The Scottish side is covered by the Scottish West Coast Regional Inshore Fisheries Group, RIFG.) In 2018 the *Solway Protector*, thirty years old, was replaced by the twin-hulled *North Western Protector* which has crew accommodation and lab space.

12. The Irish Sea is a complicated area as regards jurisdiction over fishing rights and protection of the marine ecosystem: it is bordered by five countries – Northern Ireland, Ireland, Scotland, England and Wales, which – with the Isle of Man – all have different rules. In relation to Scottish waters, for example, the statutory body Marine Scotland oversees Scottish inshore and offshore fishing, with the non-statutory Regional Inshore Fisheries Groups overseeing fishing out to 6 nautical miles (nm). For English waters, the statutory body Marine Management Organisation (MMO) oversees inshore and offshore fishing off the English coast up to 12nm, and from 12nm to the edge of the current Exclusive Economic Zone where relevant; the statutory Inshore Fisheries Conservation Authorities (IFCAs) oversee fishing up to 6nm and can impose bye-laws. The Solway Firth is included in the (English) North-West IFCA group and the (Scottish) West Coast RIFG. Marine Plans are in process of being developed by the devolved authorities, plus the Isle of Man; the MMO's NorthWest Marine Plan, which includes the English side of the Solway, should be finalised in 2021; Scotland's National Marine Plan was approved in 2015. A Solway Regional Marine Plan (for the integrated *whole* of the Solway Firth) is to be hoped for! There are additional complications, like the Crown Estate owning parts of the foreshore and seabed along the Firth . . . River basins also impact on the Solway Firth, and the Scottish Environmental Protection Authority (SEPA) and the English Environment Agency (EA) work together on the trans-boundary Solway Tweed River Basin Management Plan. The EU-funded project SIMCelt (Supporting Implementation of Maritime Spatial Planning in the Celtic Seas), which ended in 2018, sought to disentangle the various European legal requirements, and national organisational bodies and their differing metrics, to show a way forward: the Solway Firth, is an example of planning across borders. See Case Study#3 on Trans-Boundary Cooperation https://maritime-spatial-planning. ec.europa.eu/projects/supporting-implementation-maritime-spatial-planning

-celtic-seas. (My grateful thanks to Clair McFarlan, Partnership Manager of the Solway Firth Partnership, for helping me get to grips with this – any remaining mistakes are mine!)

13. One nautical mile is 1.15 statute miles (1,825 metres); in terms of speed, 1 knot is 1nm/hour (1.85 kilometres/hour).
14. Norman Nicholson, *Cumberland and Westmorland* (Robert Hale Ltd, 1949).
15. Kathleen Jamie, *Sightlines* (Sort of Books, 2012).
16. Edmund Gosse, *The Life of Philip Henry Gosse FRS* (Kegan Paul, 1890).
17. Philip Henry Gosse, *The Aquarium* (J. van Voorst, 1856).
18. Richard Mabey, 'An Owl for Winter', *The Clearing*, 2017. Online at https://www.littletoller.co.uk/the-clearing/owl-winter-richard-mabey/.

1. INVERTEBRATES ON THE EDGES

1. Rachel Carson, *The Edge of the Sea* (Staples Press, 1955).
2. For more details of the Solway Junction Railway and viaduct, see the *Crossing the Moss* website, Chapter 10 'Disaster and Demolition': https://crossingthemoss.wordpress.com/2017/04/02/10-disaster-and-demolition/.
3. *Carlisle Journal*, 1 February 1881.
4. *Carlisle Journal*, 4 February 1881.
5. P.H. Gosse, *Actinologia Britannica: A History of the British Sea - anemones and Corals* (Van Voorst, 1860). Online at https://archive.org/details/actinologiabrita00goss – see the Dahlia Wartlet *Anemone* p. 215.
6. C.B. Macfarlane, D. Drolet, M.A. Barbeau, D.J. Hamilton and J. Ollerhead, 'Dispersal of Marine Benthic Invertebrates Through Ice Rafting', *Ecology* 94 (2013): 250–6.
7. Leopold (1822–95) and his son Rudolf Blaschka (1857–1939) lived in Dresden; they started their glass-blowing business making artificial eyes, but were persuaded to make models of soft-bodied invertebrate animals like sea anemones and jellyfish, squid and sea-slugs. They made the famous Glass Flower Collection that is held at Harvard University, Boston. Blaschka models of invertebrates are held in several places, including the National Museum of Scotland, the Natural History Museum in London, Dublin's 'Dead Zoo' and the National Museum of Wales, Cardiff.
8. R. Buchsbaum, *Animals without Backbones* (University of Chicago Press, 1938), p. 3.
9. Ibid., p. 5.
10. Ann Lingard, *Seaside Pleasures* (Littoralis Press, 2003).
11. Dr Larissa Naylor's research on *Sabellaria* was carried out during her PhD, see for example L.A. Naylor and H.A. Viles, 'A temperate reef builder:

an evaluation of the growth, morphology and composition of *Sabellaria alveolata* (L.) colonies on carbonate platforms in South Wales', *Geological Society, Special Publications* 178 (2000), pp. 9–19. She is now a Reader in Biogeomorphology at the School of Geographical and Earth Sciences, University of Glasgow.

12. Rachel Carson, *The Edge of the Sea*, p. 28.

13. Caerlaverock has two nature reserves, both of which are very important for birds of the merse and mudflats, especially migrants. The western reserve is managed by Scottish Natural Heritage (rebranded NatureScot); the other is managed by the Wildfowl and Wetland Trust (WWT). My guides, Adam Murphy and Andy Over, are Nature Reserve Officers for SNH.

14. P.S. Meadows and Alison Reid, 'The behaviour of *Corophium volutator* (Crustacea: Amphipoda)', *Journal of Zoology* 150 (1966), pp. 387–99.

15. 'Sandpipers and sediments', *Fundy Issues*, Autumn 1996, no. 3.

16. Sculptor Rebecca Nassauer and I submitted a project, 'Games of Chance: The Parasite's Roulette Wheel', to the Wellcome Trust's SciArt 2000 competition. Our project was shortlisted but didn't win; we had a very stimulating few months discussing the ideas and creating the exhibits even so. The winners were artist Dorothy Cross and her scientist brother Tom with a project about the Irish zoologist Maude De Lap and her work on medusae (jellyfish) on the Irish island of Valentia – so invertebrates won in the end.

17. P.H. Gosse, *Natural History: The Mollusca* (J. van Voorst, 1854), p. 39.

18. Ted Hughes, 'Whelk', from *The Mermaid's Purse* (Faber & Faber, 1999).

2. 'CHANGEABLE DEPTHS'

1. Sea level with reference to Chart Datum and Ordnance Datum: Chart Datum (CD) is the plane below which all depths are published on a navigational chart, and the plane of reference for all tidal heights. If you add the tidal height to the charted depth, you get the true depth – this is especially important when entering ports. As explained on the EasyTide website, to determine how tide levels vary along any stretch of coast, it's necessary to refer all levels to a common horizontal plane. CD varies from place to place, so is not helpful in this respect. Ordnance Datum (Newlyn) is such a horizontal plane – it's the datum of the land-levelling system, and should be used whenever comparisons of absolute height are required. (Hence it is used as the reference level in research on coastal geomorphology.)

2. The Solway Coastwise and SCAPE report on Archaeological Surveys of *Fauna* and *Monreith* can be downloaded from https://scapetrust.org/project_category/project-reports/.

3. Brian Blake, *The Solway Firth* (Robert Hale Ltd, 1955), p. 29.

4. Peter Messenger, *Repairing clay buildings and Cumbria's clay dabbins*: https://www.buildingconservation.com/articles/clay-buildings/clay-buildings.htm.

5. Nina Jennings, *Clay Dabbins: Vernacular Buildings of the Solway Plain* (Cumberland and Westmorland Antiquarian and Archaeological Society, 2003).

6. Nina Jenning's obituary: https://www.theguardian.com/global/2015/mar/02/nina-jennings-obituary.

7. A.J. Scott, *Solway Country: Land, Life and Livelihood in the Western Border Region of England and Scotland* (Cambridge Scholars Publishing, 2015), p. 40.

8. See the entry on Blast Furnaces coke-fired, in the Industrial History of Cumbria: https://www.cumbria-industries.org.uk/a-z-of-industries/iron-and-steel/#Coke, and find the link to 'Slag' on Russell Barnes' website: http://www.users.globalnet.co.uk/%7Erwbarnes/workgton/slag.htm.

9. C. Rose, P. Dade and J. Scott, 'Qualitative and Quantitative Research into Public Engagement with the Undersea Landscape in England', *Natural England Research Reports* (NERR019, 2008).

10. William Huddart, *Unpathed Waters: The Life and Times of Captain Joseph Huddart FRS 1741–1816* (Quiller Press, 1989).

11. Sediment survey around Robin Rigg and other wind farms; Collaborative Offshore Wind Research into the Environment (COWRIE); ABPmer Ltd et al., 'A Further Review of Sediment Monitoring Data (project reference ScourSed-09': https://www.researchgate.net/publication/267221965_A_Further_Review_of_Sediment_Monitoring_Data, 2010).

12. Dr Jane Lancaster is Principal Offshore Ecologist for Newcastle-based Natural Power (www.naturalpower.com), and oversaw the underwater survey of the turbine bases.

13. 'Sandpipers and sediments', *Fundy Issues*, Autumn 1996, no. 3.

14. A.J. Einfeldt and J.A. Addison, 'Anthropocene invasion of an ecosystem engineer: resolving the history of *Corophium volutator* (Amphipoda: Corophiidae) in the North Atlantic', *Biological Journal of the Linnean Society* 115 (2015), pp. 288–304. Online at https://academic.oup.com/biolinnean/article/115/2/288/2236003.

15. Description and photos of the Richibucto ballast dumps, available online at https://www.historicplaces.ca/en/rep-reg/place-lieu.aspx?id=16346.

16. G.H. Wolff, H.H. Thoen, J. Marshall, M.E. Sayre and N.J. Strausfeld, 'An insect-like mushroom body in a crustacean brain', *eLife Sciences* (2017). Online at https://elifesciences.org/articles/29889.

3. SHIPS AND SEAWEEDS

1. Film-maker and photographer Julia Parks' website is http://juliaparks.co.uk.

2. The Marine Traffic website, marinetraffic.com, shows ship movements in real time, with details of size, cargo, route etc.

3. Francis Pryor, *The Making of the British Landscape* (Allen Lane, 2010).

4. The Irish Sea is bordered by five countries – Northern Ireland, Ireland, Scotland, England and Wales, and includes the Isle of Man – all of which have different rules (for more information see Introduction, Note 12).

5. James Irving Hawkins, *The Heritage of the Solway Firth* (Friends of Annandale and Eskdale Museums, 2006).

6. Peter Ostle, *Allonby: a short history and guide* (P3 Publications, 2014). Peter Ostle's online blog is *Solway Past and Present*. He has also co-edited, with Stephen Wright, *A Century around Silloth* (P3 Publications, 2012).

7. Joseph Huddart. See also Chapter 8.

8. The SCAPE Trust from St Andrews University, and its Scotland's Coastal Heritage at Risk Project have been working with Nic Coombey of Solway Coastwise, a project of the Solway Firth Partnership.

9. George Davidson, *The Navigation of the River Dee*. This fascinating account of piloting ships up the River Dee was written by the late George Davidson, and is reproduced on the *Old Kirkcudbright* website. My thanks to Jim Bell, who runs the website, for ready permission to use some of George's words: www.old-kirkcudbright.net/extracts-articles/books/tankers/ (2001).

10. Video: the *CEG Universe* docks at Kirkcudbright to collect crushed scallop shells from West Coast Sea products (https://www.youtube.com/watch?v=rB0oU5XXbTw).

11. See George Davidson at www.old-kirkcudbright.net/extracts-articles/books/tankers/, and Note 9 above.

12. *Marine INNS in the Solway*, Solway Firth Partnership's report. Available to download from www.solwayfirthpartnership.co.uk/environment/invasive-non-native-species/.

13. Hayden Hurst, 'Monitoring marine invasive non-native species in marinas in North-West England', report for the North-West Wildlife Trusts (2016).

14. William Harvey, *Phycologia Britannica, or, A History of British Sea-weeds* (Reeve Brothers, 1846–51). Available online from the Biodiversity Library: https://www.biodiversitylibrary.org/item/101639#page/5/mode/1up.

15. Anna Atkins, *Photographs of British Algae: Cyanotype Impressions* (1843–53). Available to view online on the British Library and New York Public Library websites.

16. Julia Parks' project to develop film using seaweed extracts is on her website at http://juliaparks.co.uk/bbc-bladderwrack/.

4. MARSHES AND MERSES

1. The cord-grass, *Spartina,* is classified as a Marine INNS – a Marine Invasive Non-Native Species (see Chapter 3).
2. There are two nature reserves at Caerlaverock; the one to the west nearest to the River Nith is managed by Scottish Natural Heritage, SNH (renamed NatureScot in May 2020); the one to the east nearest to Lochar Water is managed by the Wildfowl and Wetlands Trust, WWT.
3. *Tide Islands and Shifting Sands*, produced by the Solway Firth Partnership, is one of several booklets about the coast that can be downloaded from their website: www.solwayfirthpartnership.co.uk/community/tidelines-and-other -publications/ (2018).
4. Norman Nicholson, *Cumberland and Westmorland* (Robert Hale Ltd, 1949).
5. The tadpole shrimp, *Triops cancriformis,* is classified as endangered, listed as a priority Biodiversity Action Plan (BAP) species and specially protected under Schedule 5 of the Wildlife and Countryside Act 1981.
6. F. Balfour-Browne, 'Re-discovery of *Apus cancriformis', Nature* 162 (1948), p. 116.
7. G. Sellers, L.R. Griffin, B. Hänfling and A. Gomez, *A new molecular diagnostic tool for surveying and monitoring* Triops cancriformis *populations* (2017). Online at PeerJ 5:e3228: https://peerj.com/articles/3228/.
8. Tom Pickard, *Winter Migrants* (Carcanet Press, 2016).
9. John Young, *Robert Burns, A Man for All Seasons* (Newbattle: Scottish Cultural Press, 1996). Burns' poems are available online at www.robert-burns.org.
10. From Robert Burns' 'Elegy on Willie Nicol's Mare' (1790).
11. George Neilson, 'Annals of the Solway until AD 1307', *Transactions of the Glasgow Archaeological Society* (1899). Republished 1974 by Michael Moon at The Beckermet Bookshop, Cumbria.
12. Conservation designations of Rockcliffe Marsh: SPA (Special Protected Area); part of the Solway Coast AONB (Area of Outstanding Natural Beauty); SSSI (Site of Special Scientific Interest); SAC (Special Area of Conservation); Ramsar site (a wetland site designated as of international importance); and the area around Rockcliffe at the mouth of the Eden is a Marine Conservation Zone to protect sparling (smelt) (see Chapter 8).
13. 'Records : The fourteenth century', in *Register and Records of Holm Cultram,* ed. Francis Grainger and W.G. Collingwood (Kendal, 1929), pp. 136–48. Online at www.british-history.ac.uk/n-westmorland-records/vol7/.

14. The Historic England listing for the Church of St John the Baptist at Newton Arlosh is online at https://historicengland.org.uk/listing/the-list/list-entry /1212611.

15. John Curwen, 'The Fortified Church of St John the Baptist, Newton Arlosh', *Transactions of the Cumberland and Westmorland Antiquarian and Archaeological Society*, Series 2, 13 (1913), pp. 113–21.

16. Jenny Uglow, *The Pinecone: the story of Sarah Losh, forgotten Romantic heroine – antiquarian, architect and visionary* (Faber & Faber Ltd, 2012).

17. A. Cleaver and L. Park, *The Lonnings of Cumbria: An Exploration of the Quiet Lanes and Footpaths of Cumbria* (Create Space Publishing, 2015).

18. A.J.L. Winchester and E.A. Straughton, 'Stints and sustainability: managing stock levels on common land in England, c. 1600–2006', *Agricultural History Review* 1:58 (2010), pp. 30–48. Online at www.bahs.org.uk/AGHR/ ARTICLES/58_1_2_Winchester_Straughton.pdf.

19. Internal report by Jacqui Kay, the summer warden for Rockcliffe Marsh, for Cumbria Wildlife Trust, Summer 2001.

20. The Sediment Ecology Research Group (The 'Mud Lab') is based at the Scottish Ocean Institute, University of St Andrews (see Chapter 7).

21. Walter Newall, in 'First and Second Reports of the Commissioners, Great Britain, Tidal Harbour Commission' (1847).

22. See Kingholme Quay on the *Ports and Harbours of the UK* website: http:// ports.org.uk/ port.asp?id=819.

5. PEAT

1. Richard Lindsay, *Peat bogs and Carbon: a critical synthesis*, for RSPB Scotland (2010) – what he calls his 'Big Bogs Report'. Only available online, at https: //www.rspb.org.uk/Images/Peatbogs_and_carbon_tcm9–255200.pdf.

2. Thomas Pennant, *A Tour in Scotland and Voyage to the Hebrides in 1772*. Vol. 1 (1774).

3. 'Remembering the Solway': an ongoing oral history project organised by Solway Coast AONB (www.solwaycoastaonb.org.uk), originally by the Solway Wetlands Landscape Partnership. Video at https://vimeo.com/223637939.

4. Robin Crawford, *Into the Peatlands* (Birlinn, 2018).

5. Anne Campbell, *Rathad an Isein, The Bird's Road: a Lewis moorland glossary* (Glasgow: Faram Publications, 2013). See also www.annecampbellart.co.uk.

6. Brian Blake, *The Solway Firth* (Robert Hale Ltd, 1955), p. 201.

7. S.E. Hollingworth, 'The Glaciation of Western Edenside and Adjoining Areas and the Drumlins of Edenside and the Solway Basin', *Quarterly Journal of the Geological Society* 87 (1931), pp. 281–359.

8. Ann Lingard and James Smith, *Crossing the Moss: Bowness Common and the Solway Junction Railway*, website www.crossingthemoss.wordpress.com (2017).

9. From *The Scotsman*, quoted in *Whitehaven News*, 18 October 1866.

10. *Carlisle Journal*, 31 March 1865.

11. Terry Coleman, *The Railway Navvies* (Penguin, 1965).

12. Richard Fortey, *The Wood for the Trees: A Long View of Nature from a Small Wood* (Collins, 2016).

6. RED

1. Charles Dickens and Wilkie Collins, *The Lazy Tour of Two Idle Apprentices* (1857). Online at https://www.gutenberg.org/ebooks/888.

2. 'Obituary for Professor Robert Harkness, FRS, FGS', *Geological Magazine*, 5: 12 (1878), pp. 574–6. Online at https://www.cambridge.org/core/journals/geological-magazine.

3. R. Harkness, 'On the New Red Sandstone of the Southern Portion of the Vale of the Nith', *Quarterly Journal of the Geological Society*, 6 (1850), pp. 389–99. Online at https://doi.org/10.1144/GSL.JGS.1850.006.01-02.41; R.I. Murchison and R. Harkness, 'On the Permian Rocks of the North-West of England, and their Extension into Scotland', *Quarterly Journal of the Geological Society*, 20 (1864), pp. 144–65.

4. Photographs of different types of red sandstones, and quarries, can be found on the British Geological Survey website http://geoscenic.bgs.ac.uk/asset-bank/action/viewHome (search for e.g. St Bees' stone).

5. From 'The Seven Rocks: St Bees Sandstone' (1954), by Norman Nicholson in *Collected Poems*, ed. Neil Curry (Faber & Faber, 1994).

6. Conservation designations around St Bees and Birkham's Quarry: the St Bees Head Special Site of Scientific Interest; the Cumbria Coast Marine Conservation Zone; the Heritage Coast; the 'Colourful Coast'; Local Geo-Conservation Site; the Coast to Coast path; and the England Coast Path National Trail.

7. Henry Duncan, 'An Account of the Tracks and Footmarks of Animals found impressed on Sandstone in the Quarry of Corncockle Muir, in Dumfriesshire', *Earth and Environmental Science Transactions, Royal Society of Edinburgh*, 11 (1828), pp. 194–209.

8. John Murray IV, *John Murray III, 1808–1892: A Brief Memoir* (John Murray, 1919), p. 7.

9. *The Annals and Magazine of Natural History: including Zoology, Botany and Geology*, Vol. 6, 2nd series: various papers and contributions by Robert Harkness and William Jardine (1850).

10. M. Hyde and N. Pevsner, *The Buildings of England. Cumbria: Cumberland, Westmorland and Furness* (New Haven and London: Yale University Press, 2010).

11. Judy McKay – personal communication and family documents.

12. James Irving Hawkins, *The Sandstone Heritage of Dumfriesshire*, 2nd ed. (The Friends of Annadale and Eskdale Museum, 2014).

13. Chris Wadsworth, *The Man Who Couldn't Stop Drawing: The Extraordinary Life of Percy Kelly* (Studio Publications, 2011).

14. David A. Cross, *Cumbrian Brothers, Letters from Percy Kelly to Norman Nicholson* (Fell Foot Press, 2007).

15. 'Houses of Cistercian monks: The abbey of Holmcultram', in *A History of the County of Cumberland*, vol. 2, ed. J. Wilson (Victoria County History, 1905), pp. 162–73. Online at *British History Online*, http://www.british-history. ac.uk/vch/cumb/vol2/pp162-173.

16. Rory Stewart, *The Marches* (Penguin Random House, 2016).

17. Geoff Brown, *Herdwicks* (Hayloft Publishing, 2009).

18. Dianna Bowles, Amanda Carson, Peter Isaac, 'Genetic Distinctiveness of the Herdwick Sheep Breed and Two Other Locally Adapted Hill Breeds of the UK', *PLoS ONE* 9(1) e87823, 29 January 2014: https://doi.org/10.1371 /journal.pone.0087823; and D. Bowles 'Recent advances in understanding the genetic resources of sheep breeds locally-adapted to the UK uplands: opportunities they offer for sustainable productivity', *Frontiers in Genetics*, 6:24 (2015): https://www.frontiersin.org/articles/10.3389/fgene.2015.00024 /full.

19. Nancy Bazilchuk, 'The sheep that launched 1000 ships', *New Scientist*, 24 July 2004, pp. 52–3.

20. Gillian Eadie, 'Barrowmouth and Saltom Bay', chapter 5, part 2 of the *North West Rapid Coastal Zone Assessment (NWRCZA) Phase 2 Project*, an online research report from Historic England: https://research.historicengland.org. uk/Report.aspx?i=15775.

21. *Scotsman* quotes from *Whitehaven News*, 18 October 1866, quoted in Ann Lingard and James Smith, *Crossing the Moss: Bowness Common and the Solway Junction Railway*, website www.crossingthemoss.wordpress.com (2017).

22. *Whitehaven News*, 11 February 1869.

7. MUD LIFE

1. Alison Critchlow's website is at www.alisoncritchlow.co.uk/.

2. Norman Nicholson, 'From Walney Island' in his collection *The Pot Geranium* (1954), republished in *Collected Poems*, ed. Neil Curry (Faber & Faber, 1994).

3. M. Consalvey, D. Paterson and G. Underwood, 'The ups and downs of life in a benthic biofilm: migration of benthic diatoms', *Diatom Research* 19 (2004), pp. 181–202.

4. Professor David M. Paterson FMBA is Executive Director of MASTS, the Marine Alliance for Science and Technology for Scotland, and Director of the Sediment Ecology Research Group (SERG), https://serg.wp.st-andrews. ac.uk/, also known as the 'Mud Lab', at the Scottish Ocean Institute, St Andrews University.

5. R.J. Aspden, S. Vardy and D.M. Paterson, 'Saltmarsh microbial ecology: microbes, benthic mat and sediment movement', *The Ecogeomorphology of Tidal Marshes, Coastal and Estuarine Studies* 59 (2004), pp. 15–136.

6. Rachel Hale, R. Boardman, M.N. Mavrogordato, I. Sinclair, T.J. Tolhurst and M. Solan, 'High-resolution computed tomography reconstructions of invertebrate burrow systems', *Harvard Dataverse* (2015), http://dx.doi.org/10.7910/DVN/4XNRE3; and Rachel Hale 'Mixed assemblage: *Corophium volutator*, *Hediste diversicolor* and *Hydrobia ulvae*: *Hediste diversicolor* (worm) *Corophium volutator* (shrimp) and *Hydrobia* (snail) burrows', videos at https://www.youtube.com/watch?v=fB89Si1N7jk (2014). Dr Rachel Hale is currently a marine ecologist at the National Institute of Water and Atmospheric Research, Nelson, New Zealand. The work on *Corophium* was carried out when she was at the University of East Anglia and the University of Southampton.

7. K. Kronenberger, P.G. Moore, K. Halcrow and F. Vollrath, 'Spinning a Marine Silk for the Purpose of Tube-Building', *Journal of Crustacean Biology* 32 (2012), pp. 191–202.

8. For Marine INNS (Invasive Non-Native Species), see also Chapter 3.

9. A. Blight and D.M. Paterson, 'Impact of the non-indigenous Chinese mitten crab on sedimentary ecosystems' (November 2014): https://www.marine-vectors.eu/Core_pages/Impact_of_the_non-indigenous_Chinese_mitten_crab_(_em_Eriocheir_sinensis_em_)_on_sedimentary_ecosys.

10. *The Topographical, Statistical, and Historical Gazetteer of Scotland*, vol. 2, I–Z (Fullarton and Co., 1848). Available to browse online at https://digital.nls. uk/gazetteers-of-scotland-1803-1901/archive/97491772.

11. Cumbria Sea Fisheries Committee was replaced by the North Western Inshore Fisheries and Conservation Association (NWIFCA) in 2011. Jane Lancaster is now Principal Offshore Ecologist at Natural Power, see Note 12, Chapter 2.

12. Helen Scales, *Spirals in Time: The Secret Life and Curious Afterlife of Seashells* (Bloomsbury, 2015).

13. Edward Posnett, 'Sea silk: the world's most exclusive textile is being auctioned this week', *The Guardian*, 12 November 2019, https://www.

theguardian.com/fashion/2019/nov/12/sea-silk-byssus-auction-textile
-mollusks.

14. The last Scottish cockle survey was carried out in 2015. SeaScope Fisheries
Research Ltd (in partnership with Fruits of the Sea, University of Glasgow
and Marine Scotland Science) will undertake a new survey in 2023/4; see
https://www.solwayfirthpartnership.co.uk/scottish-solway-cockle-survey-
taking-place/.

15. P.H. Gosse, *The Aquarium : An Unveiling of the Wonders of the Deep Sea* (J. van
Voorst, 1856).

16. Lionel Playford, some weblinks: https://artandweather.blogspot.com/; Sky Gathering Lundy: https://cloudappreciationsociety.org/sky-gathering-livestream/.

17. SciArt and Rebecca Nassauer: see Note 16, Chapter 1.

18. I talk about 'squat' and piloting a ship in Chapter 2.

19. Barry Lopez, *Arctic Dreams: Imagination and Desire in a Northern Landscape*
(Harvill Press, 1999).

8. SEAFOOD

1. OD, Ordnance Datum, is Mean Sea Level at Newlyn, Cornwall (i.e. the
same as heights on land); CD, Chart Datum, is the level of the lowest astronomical tide, used for navigation purposes. See also Note 1, Chapter 2.

2. In 2023 the Marine Conservation Zone became part of the Highly Protected
Marine Area, HPMA, see Chapter 9, note 18.

3. The Marine Life Information Network: https://www.marlin.ac.uk/species/
detail/1661.

4. Neil M. Gunn, *The Silver Darlings* (Faber & Faber, 1941).

5. *Plain People* (2004) and *More Plain People* (2007) were put together and
published by the Holme St Cuthbert History Group, Cumbria, and are
collections of photographs, newspaper reports, and reminiscences from
people, 'ordinary inhabitants', of the Solway Plain.

6. William Walker, Jr (ed.), *Memoirs of the Distinguished Men of Science of Great
Britain Living in the Years 1807–8* (London: W. Davy & Son, 1864).

7. William Huddart, *Unpathed Waters: Account of The Life and Times of Captain
Joseph Huddart FRS 1741–1816* (Quiller Press: 1989), p. 10.

8. Ibid., p. 181.

9. Information on archaeological surveys of the marshes around Grune Point and
Anthorn for saltpans can be found in the *North West Rapid Coastal Zone
Assessment (NWRCZA), Phase 2 Project*, Chapter 5, part 2, pp. 331–40 (available online at https://research.historicengland.org.uk/Report.aspx?i=15775);
this includes research and surveys carried out by Dr Brian Irving, former

manager of the Solway Coast AONB. Also John Martin (1988), on the website *Industrial History of Cumbria*, http://www.cumbria-industries.org.uk/salt/.

10. George Neilson, 'Annals of the Solway until AD 1307', *Transactions of the Glasgow Archaeological Society* (1899). Republished 1974 by Michael Moon at The Beckermet Bookshop, Cumbria.

11. For further information on salt production and salt pans, see: Solway Coastwise and SCAPE Report, *Coastal recording at Redkirk Point, Gretna, Auchencairn, Dalbeattie & Cairnhead Bay, Wigtownshire*, April 2019 (published by the Solway Firth Partnership at http://scharp.co.uk/shoredig-projects/solway-coastwise-coastal-survey/).

    My thanks to Nic Coombey for sharing his information on 'The Salt Pans of Annandale', in *A History of Dumfries and Galloway in 100 Documents (part 2)* (FranScript, 2013).

12. Video about Sparling on the River Cree, by Galloway Fisheries Trust: https://www.gallowayfisheriestrust.org/save-the-sparling.php.

13. There is more information about haaf-netting and other 'heritage' fisheries on the Solway at: BBC Radio 4, *Open Country* (https://www.bbc.co.uk/programmes/b05r4088) – interviews with Cumbrian haaf-netters Mark Messenger and Mark Graham (2015); *Simply the Solway,* a video directed by Jim McMichael (2011): www.youtube.com/watch?v=OGlzNTWtC7M.

    See also Notes 14 and 15.

14. The Solway Band, 'Tom goes fishing', on *The Haafnetters* CD. Words and music for the song are by founder member of the band, Terry Croucher. Website at www.solwayband.co.uk. My grateful thanks to David Stevenson of the Solway Band for sending me the CD and allowing me to quote the words of two of the songs.

15. The Annan Common Good: photos, videos and more: http://www.annan-haafnets.org/.

16. Annan Harbour Action Group, *Ebb and Flow, Annan's Fishing Heritage, stories of then and now* (2017).

17. The Solway Band, 'The Haafnetters', on *The Haafnetters* CD.

18. George Bompas, *Life of Frank Buckland* (Smith, Elder & Co, 1885), p. 180.

19. Ibid., p. 196.

20. Celia Fiennes, *Through England on a Side-saddle*. Extracts from Celia Fiennes' journal on her *Great Journey to Newcastle and to Cornwall*, in 1698. (Penguin Books, 2009).

21. Solway Coastwise and SCAPE report; see Note 10 above.

22. A. Hale, 'Fish-traps in Scotland: construction, supply, demand and destruction' (*RURALIA V conference proceedings*, January 2003). Online at http://ruralia2.ff.cuni.cz/wp-content/uploads/2018/04/16_Hale.pdf.

23. John Steinbeck, *The Log from the Sea of Cortez*, chapter 27 (Penguin Books, 1951).

24. Annan Harbour Action Group, *Ebb and Flow*.

25. George Henry Lewes, *Seaside Studies* (William Blackwood, 1858). This verbose and rather pompous book about Lewes' shore searches and research is presumably called *Seaside Studies* to indicate its seriousness of tone in comparison with P.H. Gosse's *Seaside Pleasures*. But Gosse's book is more readable and informative.

9. CHANGING TIMES

1. *RAF Silloth*, a publication by the Local History Group of the Solway University of the Third Age (U3A). See also https://sillothairfield.wordpress.com/2015/02/25/serious-crashes-of-hudson-aircraft-from-silloth-1940-42/ and the Solway Aviation Museum: www.solway-aviation-museum.co.uk/.

2. C.M. Johnson, 'The World War II Honour Roll Project for McMaster University' (online); for details about Franklin Zurbrigg see www.mcmaster.ca/ua/alumni/ww2honourroll/zurbrigg.html.

3. Ibid.

4. The Solway Military Trail. Details and downloads online at www.solway-militarytrail.co.uk.

5. Kathleen Jamie, *Surfacing* (Sort of Books, 2019).

6. David Gange, *The Frayed Atlantic Edge: a historian's journey from Shetland to the Channel* (William Collins, 2019).

7. Tom Pickard, *Winter Migrants* (Manchester: Carcanet Press, 2016).

8. Eric Begbie, *Fowler in the Wild* (David & Charles, 1987).

9. 'Go with the floe', *The Guardian*, December 2010: www.guardian.co.uk/world/picture/2010/dec/31/weather.

10. South Solway Wildfowlers Association: the wildfowling season runs from 1 September to 20 February and – with some minor variations – shooting is not allowed between 10 a.m. and 2 p.m. The geese are most abundant in January and February, and then some of the keen wildfowlers go out two or three times a week.

11. Details for guidance and downloads at the SNH Caerlaverock reserve are online at www.nature.scot/caerlaverock-nnr-wildfowling-guidance-documents; the report for the 2018/19 shooting season is also at that link. The season is from 1 October to 20 February, not including Sundays. During that five-month period, thirty-six local seasonal shooting permits and fifty-one visitor permits were issued – but 69 per cent of those visitors and 39 per cent of the locals, shot nothing.

12. Ibid.

13. Barry Lopez, *Arctic Dreams: Imagination and Desire in a Northern Landscape* (Harvill, 1999).

14. George Neilson, 'Annals of the Solway until AD 1307', *Transactions of the Glasgow Archaeological Society* (1899). Republished 1974 by Michael Moon at The Beckermet Bookshop, Cumbria.

15. *The Topographical, Statistical and Historical Gazetteer of Scotland* (Fullarton and Co., 1848). Available to browse online at https://digital.nls.uk/gazetteers-of -scotland-1803-1901/archive/97491772.

16. Brian Blake, *The Solway Firth* (Robert Hale Ltd, 1955).

17. The Upper Solway is protected from human exploitation and 'rearrange- ment' by layers of statutory – that is, legally enforceable – conservation designations. You can investigate and overlay their virtual boundaries your- self on the excellent interactive maps at MagicMap: http://magic.defra.gov. uk/MagicMap.aspx.

18. Conservation Designations along the coasts of the Upper Solway: 'Designations', 'directives', 'habitat' and hosts of unmemorable acronyms – all vitally important for the Solway's non-human inhabitants. The largest protected area is the Upper Solway Flats and Marshes at the top end of the Firth; this is a Ramsar site – designated as important wetlands under the Ramsar Convention on Wetlands, an intergovernmental, i.e. international, treaty which 'provides the framework for national action and international co -operation for the conservation and wise use of wetlands and their resources'.

    Exactly the same area is designated under EU legislation as a European Marine Site (EMS) (my thanks to the Solway Firth Partnership for allowing me to use this extract from their website www.solwayfirthpartnership.co.uk /environment/special-places/): 'A Special Protection Area (SPA) is a site designated under the Birds Directive. These sites, together with Special Areas of Conservation (SACs), are called Natura sites and they are inter- nationally important for threatened habitats and species. Natura sites form a *unique network of protected areas which stretch across Europe* [my emphasis]. The inner Solway Firth . . . is designated as an SAC and SPA and is collectively known as the Solway Firth European Marine Site. The [separate] Solway Firth SAC designation reflects the importance of the site's marine and coastal habitats including merse (saltmarsh), mudflats and reefs. The Upper Solway Flats and Marshes SPA designation recognises the large bird popula- tions that these habitats support, particularly in winter.' It is also a Site of Special Scientific Interest (SSSI) under UK statutory protection (overseen by Natural England and Scottish Natural Heritage, respectively). So it is not trivial.

Marine Conservation Zones (MCZs – known as Marine Protected Areas, MPAs, in Scotland) are at Rockcliffe and the mouth of the River Eden (the Solway MCZ); Allonby Bay MCZ, and the coast to the west of St Bees (Cumbria Coast MCZ); MCZs are designated by DEFRA and (supposedly) protected by the UK's Marine and Coastal Access Act, which in turn was set up in response to the European Marine Strategy Framework Directive.

In July 2023, DEFRA upgraded part of the Allonby Bay MCZ to the new status of Highly Protected Marine Area (HPMA).

The Solway Coast Area of Outstanding Natural Beauty (AONB) includes fifty kilometres of Cumbrian coastline stretching from Maryport along the dunes and saltmarshes to Rockcliffe, managed in statutory compliance with the Countryside and Rights of Way Act 2000 (CROW) and overseen by the local councils, and Natural England; the AONB incorporates SSSIs too.

As for legislative 'teeth', this varies: a way into this very complex topic is through the Marine section of the excellent website, *Law & Your Environment: The plain guide to environmental law*: www.environmentlaw.org.uk/rte.asp?id=270.

19. Tidal Lagoon Power's website and updates: http://www.tidallagoonpower. com/.

20. Charles Hendry's report on tidal power possibilities can be viewed online at https://hendryreview.files.wordpress.com/2016/08/hendry-review-final-report-english-version.pdf.

21. Tidal Electric's website: http://www.tidalelectric.com/. As at March 2020, this site is uninformative about the status of their plans.

22. Solway Energy Gateway's proposed scheme: www.solwayenergygateway. co.uk/.

23. *The Scotsman*, 1 June 1925.

24. Sea-level rise prediction map for 2030 is at www.climatecentral.org; the Met Office have also produced a report and data showing changes that can be expected as far in the future as 2300: https://www.gov.uk/government/publications/exploratory-sea-level-projections-for-the-uk-to-2300.

25. Neil M. Gunn, *The Well at the World's End* (Souvenir Press, 1951).

26. Simon Armitage, 'Nature has come back to the centre of poetry', *The Guardian*, 21 November 2019, after launching the Laurel Prize for poetry, which aims to highlight 'the challenges facing our planet'. See https://www.theguardian.com/books/2019/nov/21/simon-armitage-nature-has-come-back-to-the-centre-of-poetry.

Acknowledgements

My thanks to James Smith for the use of his aerial photos in Plate I. Thanks are also due to Kathleen Jamie for permission to reproduce an excerpt from her poem 'Basking Shark'; to Simon Armitage for the lines from 'so the peloton passed' from his forthcoming collection *New Cemetery* (© Simon Armitage) published by Faber & Faber; to David Higham Associates for permission to quote lines from Norman Nicholson's poems 'The Five Rivers', 'From Walney Island' and 'The seven rocks: St Bees' Sandstone' and to Tom Pickard and Carcanet Press for permission to quote from the poem 'Winter Migrants'. I'm also very grateful to fellow authors Mark Cocker and David Gange for their enthusiastic comments about my book which appear on the jacket.

So many people – friends and new acquaintances – have helped me discover the Solway Firth: by accompanying me on walks along the edges or even in the water, by sharing their knowledge and stories, and by pointing me in the direction of new information and other people who might help. I could not have written this book without their warm and generous help and it has been such a pleasure to share a great deal of amusement with them as well.

My friend Frank Mawby, now retired from Natural England, has always been a store of useful information and ideas, especially about bogs and birds, and I thank him for reading and commenting on the 'peat' chapter; Clair MacFarlan and Nic Coombey, of the Solway Firth Partnership, have been unfailingly helpful about the 'Scottish side', and my thanks to Nic too for reading sections of chapters and for sharing so much of his own information.

For fishy matters, David Dobson (formerly of the Cumbria Sea Fisheries Committee) for long chats and a trip on the *Solway Protector*; Tom Dias for views about haaf-netting; Mark Messenger for taking me out into the Firth to haaf-net, for lending me oversize waders, and for our wade across a 'wath' (and for the 'Special' of Solway salmon at The Highland Laddie Inn); to Danny Baxter at Silloth for an entertaining morning talking about brown shrimps; and to my kind and enthusiastic friend Ronnie Porter, for a shore-walk and conversations about named stones, shrimps, herrings, ships' keels – and so much more.

For ships, piloting and ports: Russell Oldfield, the harbour master at the port of Workington, for conversations, chats and his readiness to give me information, and for taking me out on the Firth on the tug *Derwent* to meet the *Zapadnyy*; at the port of Silloth – former harbour masters and pilots Chris Puxley and Ed Deeley, and current harbour master Tim Riley, who have all been generous with their time and information; and ABP's hydrographer based at Barrow, Chris Heppenstall, for explaining how he surveys the shipping channel and for sharing data and charts; Alan Thomson, the Harbour Development Officer of the Annan Harbour Action Group; and John Whitwell and other volunteer staff at the Maryport Maritime Museum for their enthusiastic help with regard to wrecks and paintings; to Eddie Studholme at Silloth RNLI for information and contacts, and to John Stobbart of Workington RNLI for 'ordering' me to drive the lifeboat.

For matters connected with peat and bogs: Frank Mawby for all manner of help, including sessions looking at his magnificent archival collection of documents and photos of the South Solway Mosses; to Alasdair Brock, formerly of Natural England, and Dave Blackledge, of RSPB Campfield, for sharing so much information about Bowness Common; Dr Lauren Parry of the University of Glasgow for a session peat-coring on Kirkconnell Flow, and to Kate Foster and Dave Borthwick for conversations and company; to Patrick McGoldrick and Thomas Holden for their detailed and moving memories of cutting peat by hand; and of course to my friend and collaborator on the *Crossing the Moss* project, photographer James Smith.

For matters concerning the New Red Sandstone: David Kelly, for the field trips to St Bees Head and Barrowmouth Bay, which inspired my love of the red rock and set me on the mission to find out more; sculptor Sky Higgins for her insights into the challenges and delights of working with New Red (and for her fine piece, *Phoenix in Flames*); to Judy McKay (her maiden name) and for sharing her birthday tea with me, and so many stories, letters and documents about her family's long-term involvement with quarrying, masonry and brick-making, both sides of the Firth; to my oldest (that is, longest-term) friend, and industrial archaeologist, Dr Peter Stanier, for new ways of looking at the rocks in Fleswick Bay, for ideas, and for reading and commenting on the 'sandstone' chapter; to archaeologist Mark Graham for sharing information about Holme Cultram Abbey; and to Lara Reid for giving up her time to read and comment on the 'sandstone' chapter.

It was a delight and inspiration to visit the Campfield saltmarsh with the late Norman Holton – I continue to thank him for opening my eyes to the special beauty of these places; doctoral student Guillaume Goodwin and his friends entertained me with accounts of their salt-marsh research on a freezing and windy day; my thanks to Imogen Rutter and Kevin Scott, of Cumbria Wildlife Trust, for trips and laughter out on Rockcliffe Marsh; I'm especially grateful to Giles Mounsey-Heysham, owner of Castletown Estate and Rockcliffe Marsh for trying (and failing) to teach me to ride a quad bike (fast) and for an adventurous morning spent out on the marsh and mudflats; to Dr Bart Donato, of Natural England, for a long and amusing conversation and for bringing his play dough to explain saltmarsh hydrology; to Eileen Bell of Newton Arlosh for her kindness in taking me out onto the Newton Arlosh Marsh, for tea and scones, and for explaining (some of) the complexities of stints and grazing; to Dave Blackledge again, and this time for conversations about the Campfield saltmarsh and for joining me on the mudflats to look for *Corophium*; and to Peter Norman for a long, hot and fascinating walk across Kirkconnell Merse to investigate the 'training wall' and for sharing some of his research; to Brian Hodgson, at that time secretary of the South Solway Wildfowlers Association, for dawn sojourns out on Border and Calvo Marshes; and to Adam Murphy and Andy Over, Reserve Officers at SNH's Caerlaverock Reserve, for showing me the saltmarsh and mudflats by the River Nith, and their enthusiasm – even on a bitterly cold day – for digging for *Corophium*.

For muddy matters, my thanks to Professor David Paterson at the 'Mud lab' (SERG), Scottish Ocean Institute, St Andrews, for a guided tour of the lab, conversations, research papers and a great deal of help in understanding the rôle of the microphytobenthos in mudflats and marshes; to Dr Andy Blight, lab manager at SERG, for further inform-ation, especially about the Chinese mitten crab; and to Dr Rachel Hale, now in New Zealand, for email conversations about *Corophium* burrows.

Brian Irving, former manager of the Solway Coast AONB, an enthusiast for all matters relating to the English Solway coast, shared many ideas and information with me over several years; also at the Solway Coast AONB, and while working on the Solway Wetlands Partnership Project, Chris Spencer enthused us all about building a clay dabbin house (and he and current manager Naomi Kay have always been very helpful with 'Solway' matters, and very supportive of James Smith's and my *Crossing the Moss* project); I will never forget mixing dried ox blood with soil for the dabbin house floor with Alex Gibbons, earth-building specialist; and Peter Messenger, clay dabbin 'guru', provided much helpful information.

As regards *Corophium* and other invertebrate animals that live along the Solway's shores, I am so grateful to Dr Jane Lancaster, now at Natural Power, for (early in my time here on the Solway) taking me far out to Ellison's Scaur to show me the great mussel beds, and to various other shores to look at honeycomb-worm reefs – and for a great deal of other help and information over the years; to Dr Larry Griffin, at WWT Caerlaverock Reserve, for talking to me about the rare tadpole shrimps, *Triops*, and to Dr Africa Gomez at the University of Hull for conversations about the work on eDNA of her doctoral student Graham

Acknowledgements

Sellar; to Dr Larissa Naylor, of the University of Glasgow, for information and images from her earlier work on tube-building in the honeycomb worm, *Sabellaria*; to Dr Emily Baxter of Cumbria Wildlife Trust for giving me a copy of Hayden Hurst's report on his research into Marine INNS (and to Hayden, of course, for his interesting work); to my long-time friends and former colleagues at the University Marine Biological Station, Millport, professors Jim Atkinson and Geoff Moore, for coffee and conversations, and for sharing information on burrowing Crustacea, including *Corophium*; to Dr Tony Einfeldt, for very interesting correspondence about the spread of *Corophium* populations, perhaps in ships' ballast.

Professor David Smith has been unfailingly helpful with matters relating to the geomorphology and changes in sea level in the Solway Firth.

I thank film-maker Julia Parks for shore-walks-and-talks, and explaining her novel ideas on making and developing films. Artists Alison Critchlow and Lionel Playford have been wonderful companions on the saltmarshes and muddy shores, helping me to see and think about these shore-scapes in entirely new ways.

Back in 2003 Keith Richardson, then the editor of *Cumbria Life*, encouraged me to write several pieces for the magazine, and I thank him for getting me started on my Solway wanderings and writings.

I am enormously grateful to Hugh Andrew and Andrew Simmons at Birlinn, for their good humour, encouragement and wealth of helpful suggestions as I wrote this book.

And of course my love and thanks go to my family: daughters Kate and Rachel, for their continuing support (and especially to Rachel for reading some chapters of a very early draft and providing wise advice!), and to my husband, John – without his willingness to join me in heading off to obscure places, and to accompany me across soggy bogs and down cliff-faces, those 'expeditions' would not have been half so much fun.

Index